Exercises in Organic Chemistry

Veljko Dragojlovic
Wilkes Honors College of Florida Atlantic University

Copyright © 2005, 2006, 2011 by Veljko Dragojlovic. All rights reserved.

No part of this book may be reproduced or transmitted in any form or by any means, electronic or mechanical, including photocopying, recording, or by any information storage and retrieval system, without written permission from the publisher, except for the inclusion of brief quotations in a review. Address inquiries to Veljko Dragojlovic, 7554 Nova Drive, Davie, FL 33317-7002.

Dragojlovic, Veljko

Exercises in Organic Chemistry

Preface

This book has been designed primarily as a supplement to Organic Chemistry textbook by Veljko Dragojlovic. As the order of topics in most current organic chemistry textbooks is similar, with minor adjustments, it may be suitable as supplement to most of them. Finally, it may be useful as a guide for students who already took organic chemistry and need a review in order to take standardized exams such MCAT, DAT or GRE.

The work on the next edition has already begun. You can provide comments on the style, content and ways in which this book could be improved to vdragojl@fau.edu.

Contents

Chapter 1. Review of General Chemistry and Introduction into Organic Chemistry   *1*
Chapter 2. Nomenclature and Representation of Organic Compounds   *11*
Chapter 3. Structure and Properties of Organic Molecules: Intermolecular Interactions   *21*
Chapter 4. Alkanes and Conformational Analysis   *23*
Test 1: Chapters 1-4   *29*
Chapter 5. Acids and Bases   *35*
Chapter 6. Organic Reactions   *41*
Chapter 7. Reactions of Alkanes: Free Radical Substitution   *47*
Chapter 8. Unsaturated Hydrocarbons: Alkenes and Alkynes   *53*
Test 2: Chapters 5-8   *59*
Chapter 9. Reactions of Alkenes   *65*
Chapter 10. Reactions of Alkynes   *75*
Chapter 11. Stereochemistry   *79*
Special Topic: Stereochemistry in Organic Reactions   *99*
Chapter 12. Substitution Reactions of Alkyl Halides   *107*
Test 3: Chapters 9-12   *121*
Chapter 13. Elimination Reactions of Alkyl Halides   *129*
Chapter 14. Substitution and Elimination Reactions of Alcohols and Ethers   *139*
Chapter 15. Spectroscopy   *147*
Test 4: Chapters 13-15   *163*
Chapter 16. Electron Delocalization and Resonance   *173*
Chapter 17. Dienes and Polyenes   *181*
Chapter 18. Aromaticity: Reactions of Benzene   *189*
Chapter 19. Reactions of Substituted Benzenes   *197*
Test 5: Chapters 16-19   *217*
Chapter 20. Carbonyl Compounds I: Aldehydes and Ketones   *229*
Chapter 21. Carbonyl Compounds II: Carboxylic Acids and Derivatives   *241*
Chapter 22. Carbonyl Compounds III: Substitution α to Carbonyl Groups   *257*
Chapter 23. Organic Polymers   *267*
Test 6: Chapters 20-23   *271*
Chapter 24. Natural Products   *283*
Final Exam   *287*

Exercises in Organic Chemistry

1. Review of General Chemistry and Introduction into Organic Chemistry

A good background in general chemistry is essential for successful study of organic chemistry. A review of general chemistry usually includes: structure of an atom, chemical bonding, chemical formulas, reaction intermediates, bonding in simple molecules, geometry of molecules, polar and non-polar bonds, dipole moments of molecules as well as concepts such as calculation of yield of a reaction.

To solve problems in this chapter you may need a periodic table and a list of electronegativities.

Periodic Table of the Elements

1A 1	2A 2	3B 3	4B 4	5B 5	6B 6	7B 7	8	9	10	1B 11	2B 12	3A 13	4A 14	5A 15	6A 16	7A 17	0 18
1 H 1.008																	2 He 4.003
3 Li 6.941	4 Be 9.012											5 B 10.81	6 C 12.01	7 N 14.01	8 O 16.00	9 F 19.00	10 Ne 20.18
11 Na 22.99	12 Mg 24.31											13 Al 26.98	14 Si 28.09	15 P 30.97	16 S 32.07	17 Cl 35.45	18 Ar 39.95
19 K 39.10	20 Ca 40.08	21 Sc 44.96	22 Ti 47.86	23 V 50.94	24 Cr 52.00	25 Mn 54.94	26 Fe 55.85	27 Co 58.93	28 Ni 58.69	29 Cu 63.54	30 Zn 65.39	31 Ga 69.72	32 Ge 72.61	33 As 74.92	34 Se 78.96	35 Br 79.90	36 Kr 83.80
37 Rb 85.47	38 Sr 87.62	39 Y 88.91	40 Zr 91.22	41 Nb 92.90	42 Mo 95.94	43 Tc (98)	44 Ru 101.07	45 Rh 102.9	46 Pd 106.4	47 Ag 107.9	48 Cd 112.4	49 In 114.8	50 Sn 118.7	51 Sb 121.8	52 Te 127.6	53 I 126.9	54 Xe 131.3
55 Cs 132.9	56 Ba 137.32	57 *La 138.9	72 Hf 178.5	73 Ta 180.94	74 W 183.84	75 Re 186.2	76 Os 190.23	77 Ir 192.2	78 Pt 195.1	79 Au 197.0	80 Hg 200.6	81 Tl 204.4	82 Pb 207.2	83 Bi 209	84 Po (209)	85 At (210)	86 Rn (222)
87 Fr (223)	88 Ra (226)	89 **Ac (227)	104 Rf (261)	105 Db (262)	106 Sg (263)	107 Bh (262)	108 Hs (265)	109 Mt (266)	110 (269)	111 (272)	112 (277)						

Electronegativities of some elements

H
2.2

Li	Be
1.0	1.6

Na	Mg
0.9	1.3

K	Ca
0.8	1.0

B	C	N	O	F
2.0	2.6	3.0	3.4	4.0

Al	Si	P	S	Cl
1.6	1.9	2.2	2.6	3.2

Br
3.0

I
2.7

Veljko Dragojlovic

Structure of atoms. Chemical Bonds

Questions 1-4.

1. How many valence electrons does a phosphorus atom have?

2. Classify each of the following compounds as ionic or covalent. If one of the bonds in a molecule is ionic, the compound is classified as ionic. Only if all the bonds are covalent, the compound is considered to be covalent.

$$CsCl \quad CaCl_2 \quad NO_2 \quad CH_3MgBr \quad HClO$$

3. In which of the following compounds does hydrogen have a partial positive charge?

   a) KH
   b) $HClO_4$
   c) SiH_4
   d) BH_3

4. Use either signs δ+ and δ- , or arrows, to indicate polarity of each bond.

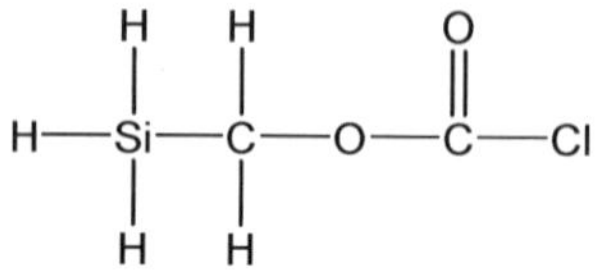

$$\begin{array}{c} H \quad H \quad\quad\quad O \\ | \quad\; | \quad\quad\quad \| \\ H-Si-C-O-C-Cl \\ | \quad\; | \\ H \quad H \end{array}$$

Exercises in Organic Chemistry

Answers to questions 1-4.

1. Phosphorus is in the group 15 and, therefore, it has 5 valence electrons.

2. Use table of electronegativities to answer this question. If difference between the electronegativities of any two elements that form a bond is 1.7, or greater, than the bond is ionic. Otherwise, it is covalent. Therefore: CsCl – ionic; $CaCl_2$ – ionic; NO_2 – covalent; CH_3MgBr – ionic (Mg–Br bond); HClO – covalent.

3. The correct answer is b). In the other three compounds, hydrogen atom has a partial negative charge.

4. It is easier to use arrows. Due to a large number of bonds, signs δ+ and δ- clutter the drawing.

However, if you prefer δ+ and δ- signs:

Formal Charges

Questions 5-15.

5. What is the formal charge on the oxygen atom?

    H—O

6. Does this help?

    H—Ö·

7. What are the formal charges on these atoms?

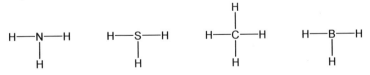

8. Can you answer the question now?

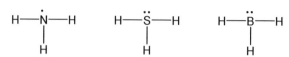

9. Why you did not need any information about electrons on the carbon atom shown above?

10. In the molecule shown below, which atom carries a negative charge?

11. How many free electrons are there on the central atoms shown below?

12. Can you answer that now?

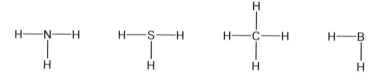

13. How about the number of free electrons on these atoms?

H—N—H   H—S—H   H—C—H   H—B
  | | | |
  H   H   H   H
             (with H above C)

14. Can you answer the question now?

H—N—H   H—S⁻—H   H—B⁺
  | | |
  H   H   H
no charge

15. Why you did not need any information about the charge on carbon to answer the question 13?

Answers to questions 5-15.

5. One cannot tell without knowing the number of free electrons on the oxygen atom.

6. Yes, it does. Now that we know the number of free electrons on oxygen we can calculate its formal charge:

4

$$\text{Formal Charge} = \begin{pmatrix} \text{number of valence} \\ \text{electrons} \end{pmatrix} - \begin{pmatrix} \text{number of} \\ \text{assigned electrons} \end{pmatrix}$$

Note that oxygen atom has one bond, two free electron pairs and one free unpaired electron. Therefore, a total number of assigned electrons on the oxygen atom is a sum of:

½ electrons from oxygen's bonds + electrons from free electron pairs + unpaired electrons

or

number of assigned electrons = 1 x 1 + 2 x 2 + 1 x 1

and the formal charge is:

formal charge = 6 – (1 x 1 + 2 x 2 + 1 x 1) = 0

7. Carbon atom has a formal charge of zero. One cannot tell about the rest.

8. Formal charges are:
nitrogen +1, sulfur +1 and boron -2.

9. In the formula CH_4, carbon atom has four bonds and therefore at least eight valence electrons. Since those electrons complete carbon's octet, it cannot possibly have any more electrons and all of its electrons are accounted for.

10. A carbon atom caries the negative charge:

$$H_3C\text{—}\overset{\ominus}{\underset{..}{C}}H\text{—}\overset{..}{\underset{..}{N}}\text{=}\overset{\oplus}{N}\text{:}$$

11. One cannot tell unless the charges are shown.

12. Now one can calculate the number of electrons. First, from the equation we encountered earlier (question 6), one can calculate the number of assigned electrons.

$$\text{Formal Charge} = \begin{pmatrix} \text{number of valence} \\ \text{electrons} \end{pmatrix} - \begin{pmatrix} \text{number of} \\ \text{assigned electrons} \end{pmatrix}$$

Thus, for the oxygen atom it is:
-1 = 6 – number of assigned electrons, or 7.

For the nitrogen atom it is:
+1 = 5 – number of assigned electrons, or 4.

$$\text{number of free electrons} = \begin{pmatrix} \text{total number of} \\ \text{assigned electrons} \end{pmatrix} - \begin{pmatrix} \text{number of assigned} \\ \text{electrons from bonds} \end{pmatrix}$$

From the above equation, it follows that the oxygen atom has 6 and the nitrogen atom 2 free electrons. Therefore, complete Lewis formulas are:

$$H-\ddot{\underset{..}{O}}:^{\ominus} \qquad H-\overset{..}{\underset{|}{N}}{}^{\oplus}$$
$$\phantom{H-\ddot{\underset{..}{O}}:^{\ominus} \qquad H-}H$$

Note that we cannot tell whether the two electrons on the nitrogen atom are paired up, or not.

13. Carbon has none. One cannot tell about the other atoms.

14. Now that the charges are known, one can apply procedure described above and come up with the following Lewis formulas:

$$H-\underset{|}{\overset{..}{N}}-H \qquad H-\overset{\ominus}{\underset{|}{\ddot{S}}}-H \qquad H-\overset{\oplus}{\underset{|}{B}}$$
$$H H H$$
no charge

Note that you had to be told that the nitrogen atom does not have a charge. Another thing worth nothing is that the sulfur atom has exceeded the octet. As a third row element, it is capable of accommodating more than eight electrons in its valence shell.

15. As in the question 7, the formula CH_4, carbon atom has its octet filled. Therefore, all of its electrons are represented and accounted for.

Additional Exercises: Questions 16-27

16. Calculate formal charges of the underlined atoms.

17. Identify the hybridization of each of the C, N and O atoms in the following structure.

18. How many electrons are there on the central atoms shown below?

19. Identify the hybridization of each of the C, N and O atoms in the following structure.

20. Calculate the formal charge on the sulfur atom in the molecule below.

$$\ddot{\underset{..}{S}}-H$$

a) -2
b) -1
c) 0
d) +1

21. The only element that has a negative oxidation state when combined with oxygen in a compound is:

a) hydrogen
b) chlorine
c) sulfur
d) fluorine

22. Which of the following is not a correct completion of the following statement? A heterolytic bond cleavage:

a) results in formation of ions.
b) results in one atom leaving with the electron pair.
c) results in formation of free radicals.
d) is assisted by the presence of a strong nucleophile or a strong electrophile.

23. Which of the following molecules do not have a dipole moment?

I. CH_4 II. CH_3F III. $N\equiv C\text{-}C\equiv N$ IV. CH_3Na

a) I and III
b) II and III
c) II and IV
d) III and IV

24. Ethane (0.70 moles) and bromine (0.50 moles) reacted to produce 0.35 moles of bromoethane. The reaction stochiometry is 1:1. What is the percent yield in this reaction?

   a) 70%
   b) 50%
   c) 85%
   d) 35%

25. Which of the following compounds contains the largest number of sp^2 hybridized carbon atoms?

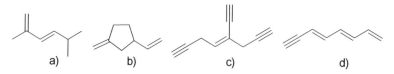

   a)   b)   c)   d)

26. Which compound, among those listed below, contains the shortest carbon-hydrogen bond?

   a)   b)   c)   d)

27. Which compound, among those listed below, contains the longest carbon-carbon bond?

   a)   b)   c)   d)

Answers to questions 16-27.

16. Formal charges:

17. Hybridization

$$\underset{sp^3}{CH_3}-\underset{\underset{\underset{sp}{N}}{\overset{\overset{sp^2}{NH}}{\|}}}{\overset{}{C}}-\underset{sp}{C}\equiv N \quad (sp^2)$$

18. Number of free electrons

- $(CH_3)_3 O^+ \cdots CH_3$ structure — **two**
- cyclopentyl-$N^-(CH_3)$ — **four**
- cyclohexyl-NH_2^+ — **two**

19. Hybridization:

$$\underset{sp^3}{HO}-\underset{sp}{C}\equiv\underset{sp}{C}-\underset{sp^2}{\overset{\overset{sp^2}{NH}}{C}}\underset{H}{\diagdown}$$

20. c); 21. d); 22. c); 23. a); 24. a); 25. d); 26. c); 27. a).

Veljko Dragojlovic

Exercises in Organic Chemistry

2. Nomenclature and Representation of Organic Compounds

Concepts to review: nomenclature of simple organic compounds, different types of structural formulas, structure of organic compounds, relationship between the structure and physical properties of organic compounds, intermolecular interactions, boiling points, melting points and solubility.

Questions 1-4.

1. Name the following compounds:

2. The following are incorrect names for organic molecules. Give correct name for each compound.

   a) 6-methylcyclohexanol
   b) 2-ethyl-5-methylhexane

3. Provide a structural formula of *trans*-2-benzylaminocyclohexanol.

4. Which of the names listed below is a correct IUPAC name?

   a) 4-ethyl-3-methylhexane
   b) 2-ethyl-2-methylhexane
   c) 2-isopropylbutane
   d) 3-ethyl-2-methylpentane

Answers to questions 1-4.

1. Naming of the first compound is a relatively simple task. One should identify and name the longest chain, then identify and name the substituents and, finally, assign smallest possible numbers to the substituents:

5-ethyl-3,4-dimethyloctane

Next two involve a little "trick" – the longest chain is not presented as a horizontal zigzag line. Once the longest chains are identified, naming should be relatively easy:

2,5-dimethyloctane

3,5-dimethylheptane

11

The final two compounds are an ether and an amine. It is easiest to use descriptive (common) names for them. IUPAC names are given in brackets.

isopropyl methyl ether
(2-methoxypropane)

isopropylethylamine
N-ethyl-2-propanamine

2. This question often confuses students. If a name is incorrect how can one know what is the actual formula of the compound? Just because the name is incorrect does not mean that one cannot associate a formula to the name. First, use the incorrect name to draw a structural formula and then name the compound correctly.

Therefore, 6-methylcyclohexanol is actually 2-methylcyclohexanol.

Correct name for "2-ethyl-5-methylhexane" is 2,5-dimethylheptane.

3. From the name it is clear that the compound is a substituted cyclohexanol. Therefore, one should start by drawing its formula.

Cyclohexanol has a substituent in on the second carbon atom and relative stereochemistry of – OH group and the substituent is *trans*.

Finally, the substituent is benzylamino group. $C_6H_5CH_2-$ is benzyl substituent and benzyl amine has a formula of $C_6H_5CH_2NH_2$. Therefore, benzylamino substituent has a formula of $C_6H_5CH_2NH-$.

Other ways to name this compound are:
trans-2-(benzylamino)cyclohexanol
trans-2-benzylamino-1-cyclohexanol
(1*R*,2*R*)-2-benzylaminocyclohexanol (note that this compound has two chirality centers – Chapter 11)

4. As when answering the question 2, you should draw the compounds and figure out if the names are correct. Thus, a) 4-ethyl-3-methylhexane is not correct because ethyl substituent should be assigned the lower number (3-ethyl-4-methylhexane); b) 2-ethyl-2-methylhexane is not correct because the longest chain is heptane (3,3-dimethylheptane); c) 2-isopropylbutane is not correct because the longest chain is pentane (2,3-dimethylpentane); d) 3-ethyl-2-methylpentane is a correct IUPAC name.

Questions 5-8.

5. What is wrong with the following formulas? Draw the correct ones.

6. Draw a complete structural formula for the following compound:

7. Draw a line formula for the following compound:

8. Which of the perspective drawings represents the same conformation as the Newman projection shown below?

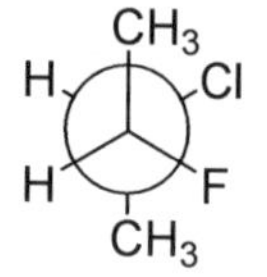

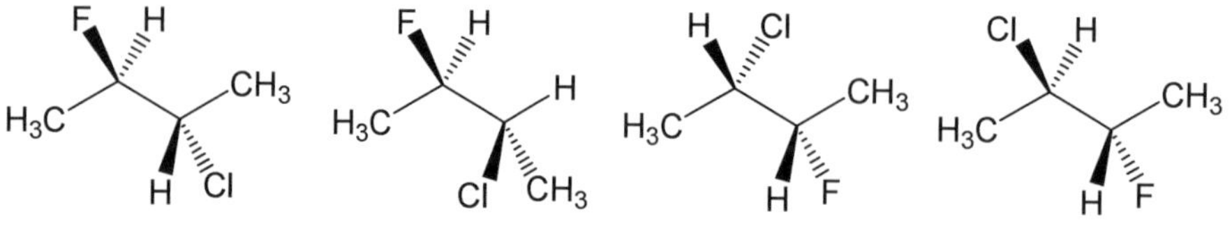

a)   b)   c)   d)

Answers to questions 5-8.

5. The formulas are incomplete. One on the left shows only the carbon atoms, while one on the right has "dangling bonds" on the carbon atoms. Note that these are not line formulas. They are simply wrong. If carbon atoms are shown, then substituents, including hydrogen atoms must be shown, too. The same applies if the bonds are represented. One cannot simply assume that at the other end of a bond is a hydrogen atom. Correct formulas are shown on the right.

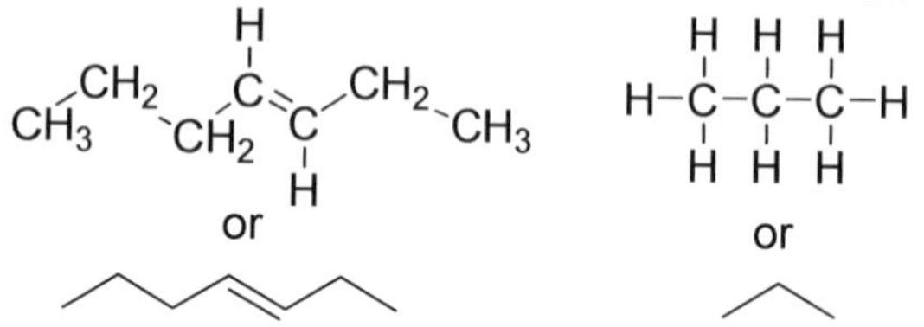

CH_3-CH_2-$CH=CH$-CH_2-CH_3 H_3C-CH_2-CH_3

or

6. The best way to do it is by first redrawing the formula and placing a carbon atom at the end of each bond. Next, add "dangling bonds" on each carbon atom to satisfy its valence of four. Finally, place hydrogen atoms at the end of each "dangling bond."

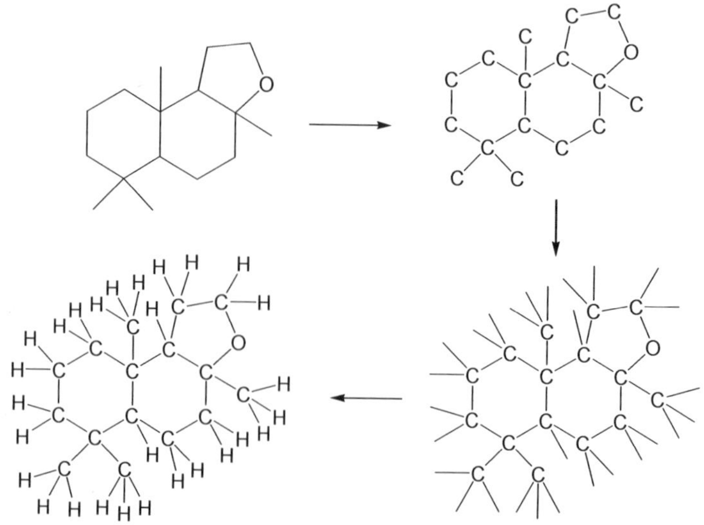

14

Exercises in Organic Chemistry

7. To draw a line formula, one should reverse the previous process. Thus, omit the hydrogens and their bonds to carbon atoms. Next omit all carbon atoms and draw the bonds only. Keep in mind that you still have to represent the heteroatoms (atoms other than carbon and hydrogen) as well as hydrogens bonded to the heteroatoms.

8. It is the formula c). If you cannot see it, you can convince yourself by making a model.

Additional Exercises: Questions 9-17.

9. Complete the following table by giving the structure of each molecule listed.

Name	Structural formula
3-methylcyclopentanol	
methyl vinyl ether	
s-butyl fluoride	
1-chloro-3-methylcyclopentane	

10. Complete the following table by providing the common (trivial) name for each molecule listed.

Name	Structural formula
	$(CH_3)_2CH\text{-}O\text{-}CH_3$
	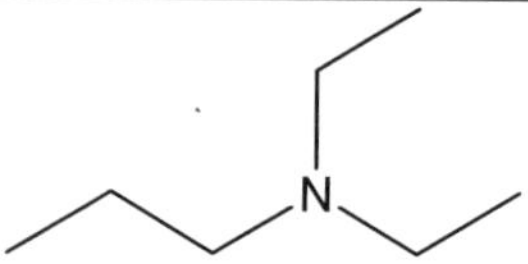
	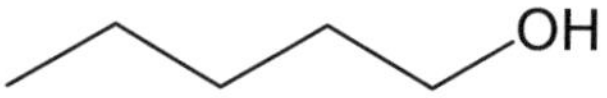
	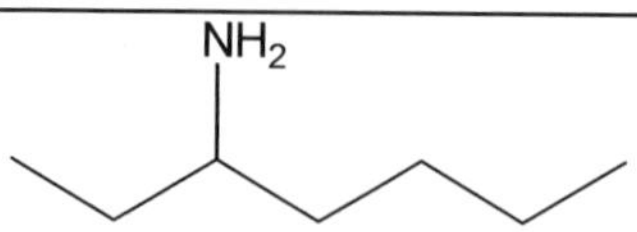
	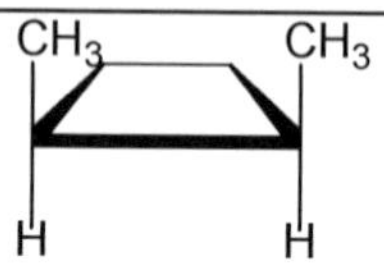

11. Correct name of the following compound is:

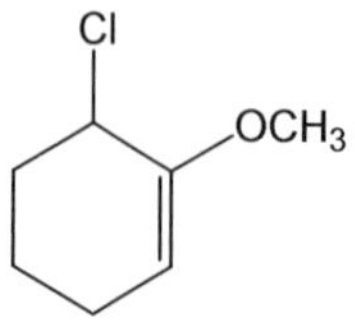

a) 3-chloro-2-methoxy-1-cyclohexene
b) 6-chloro-1-methoxy-1-cyclohexene
c) 2-chloro-1-methoxy-6-cyclohexene
d) 2-chloro-1-methoxy-1-cyclohexene

Exercises in Organic Chemistry

12. Complete the following table by providing the IUPAC name for each molecule listed.

Name	Structural formula
	(branched alkane structure)
	(cyclopentene with Cl and methyl)
	(1,2-dibromocyclopentane)
	(N-ethylaniline structure)
	(alkene with tert-butyl group)
	(enol ether alkene structure)

13. Which Newman projection represents the same conformation as the perspective drawing shown below?

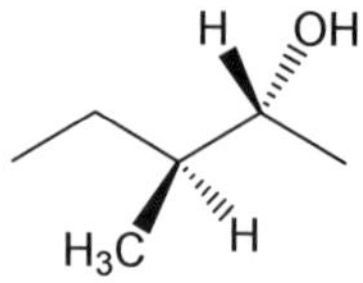

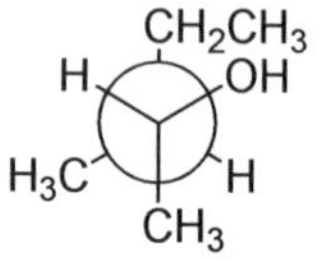

Answers to Additional Exercises: Questions 9-13.

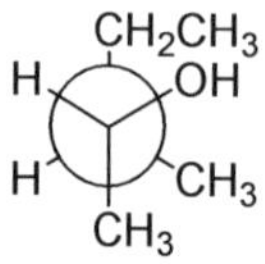

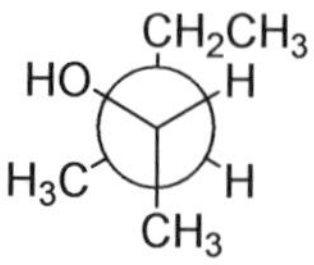

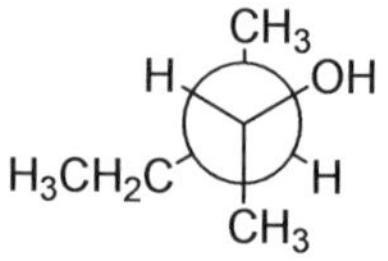

9. Complete the following table by giving the structure each molecule listed.

Name	Structural formula
3-methylcyclopentanol	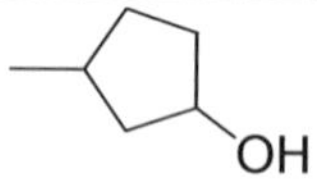 (cyclopentane with CH₃ and OH substituents)
methyl vinyl ether	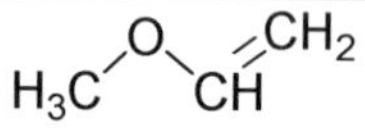 $H_3C-O-CH=CH_2$
s-butyl fluoride	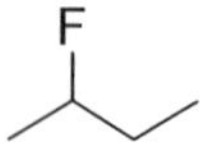 (CH₃CH₂CH(F)CH₃)
1-chloro-3-methylcyclopentane	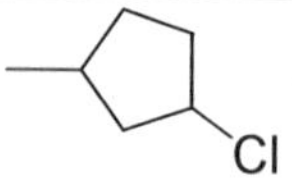 (cyclopentane with CH₃ and Cl substituents)

10. Complete the following table by providing the common (trivial) name for each molecule listed.

Name	Structural formula
isopropyl methyl ether	$(CH_3)_2CH-O-CH_3$
diethylpropylamine	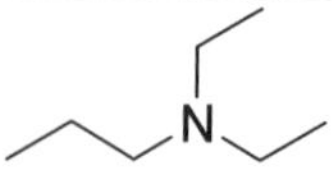 (CH₃CH₂)₂N-CH₂CH₂CH₃
amyl alcohol	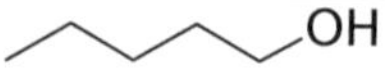 CH₃CH₂CH₂CH₂CH₂-OH

Exercises in Organic Chemistry

Name	Structural formula
3-aminoheptane or 3-heptanamine	(structure with NH₂)
cis-1,2-dimethylcyclobutane	(structure)

11. b.

12. Complete the following table by providing the IUPAC name for each molecule listed.

Name	Structural formula
2,2,5-trimethylheptane	(structure)
1-chloro-5-methyl-1-cyclopentene	(structure)
1,2-dibromocyclopentane	(structure)
N-ethylbenzenamine or *N*-ethylaniline	(structure)
(*E*)-2,2-dimethyl-3-hexene	(structure)
(*Z*)-3-methoxy-3-hexene	(structure)

13. a.

19

Veljko Dragojlovic

Exercises in Organic Chemistry

3. Structure and Properties of Organic Molecules: Intermolecular Interactions

Questions 9-11.

1. Which of the following compounds has the lowest boiling point?

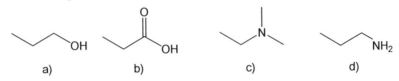

2. Explain the order of the following boiling points by indicating the type and strength of intermolecular forces which apply to each molecule.

Structure (B.P.)	Types of Intermolecular Forces
$CH_3CH_2CH_3$ (-42°C)	
CH_3CH_2Cl (12°C)	
CH_2Cl_2 (40°C)	
CH_3CH_2I (72°C)	
CH_3OH (78°C)	

3. Which of the following becomes more soluble in water upon addition of NaOH?

   a) a carboxylic acid
   b) an alkene
   c) a fluoroalkane
   d) an ether

4. Which of the following compounds are soluble in water:

   I. $CH_3CH_2CH_2CH_2CH_3$ II. CH_3NH_2 III. $CH_3CH_2CO_2H$ IV. CBr_4

   a) I and II
   b) I and III
   c) II and III
   d) II and IV

5. Which of the following compounds would you expect to be the most soluble in hexane?

   a) C_2H_5OH
   b) $C_6H_5NH_2$
   c) $C_6H_{13}I$
   d) CH_3CO_2H

6. Which of the following becomes more soluble in water upon addition of HCl?

   a) an alcohol
   b) an amine
   c) a carboxylic acid
   d) an ether

7. Which of the following compounds is a protic solvent?

   a) hexane
   b) tetrahydrofuran
   c) acetone
   d) ethanol

Answers to questions 1-7.

1. The compound c) has the lowest boiling point. All other compounds exhibit hydrogen bonding. Compound c) is a tertiary amine and exhibits only a weaker dipole-dipole interaction.

2. Completed table:

Structure (B.P.)	Types of Intermolecular Forces
$CH_3CH_2CH_3$ (-42°C)	weak van der Waals interactions
CH_3CH_2Cl (12°C)	moderate van der Waals interactions weak dipole-dipole interactions
CH_2Cl_2 (40°C)	moderate van der Waals interactions strong dipole-dipole interactions
CH_3CH_2I (72°C)	strong van der Waals interactions weak dipole-dipole interactions
CH_3CH_2OH (78°C)	weak van der Waals interactions strong dipole-dipole interactions hydrogen bonding

Note that van der Waals interactions are not always weak and that dipole-dipole interactions are not always stronger than van der Waals interactions. Large polarizable atoms (such as iodine) exhibit strong van der Waals interactions. On the other hand, presence of a single electronegative atom in an otherwise non-polar molecule may result only in weak dipole-dipole interactions.

3. a) A carboxylic acid forms a salt upon reaction with NaOH. Salts are ionic compounds and are in general water soluble due to ion-dipole interactions.

4.. c): 5. c); 6. b); 7. d.

Exercises in Organic Chemistry

4. Alkanes and Conformational Analysis

Concepts to review: conformations of open-chain compounds, conformations of cyclic hydrocarbons, types of strain (torsional, steric, angle).

Questions 1-6.

1. Different conformations of ethene are shown below. Complete the diagram by drawing relative energy levels that correspond to those conformations.

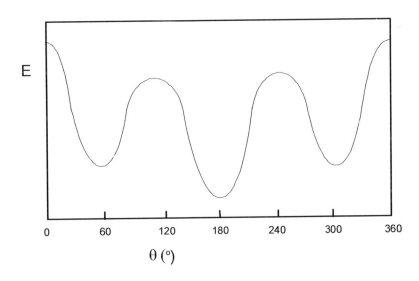

2. Diagram below shows relative energies of different conformations of butane. Complete the diagram by drawing the conformations that correspond to various points on the energy diagram.

3. Name the following conformations of butane:

4. Draw the following conformation of butane as a saw-horse formula.

5. Draw the most and the least stable conformations of 1-bromo-2-chloroethane.

6. Draw the most stable conformation of 1,2-ethanediol (CH$_2$OHCH$_2$OH).

Answers to questions 1-6.

1. Eclipsing conformations correspond to the energy maxima and staggered conformations correspond to the energy minima. Conformations between those two extremes have intermediate energies.

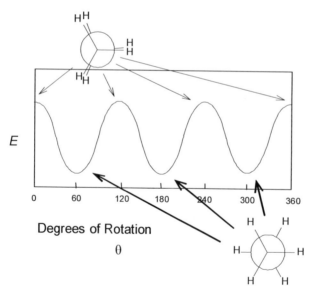

2. Completed diagram:

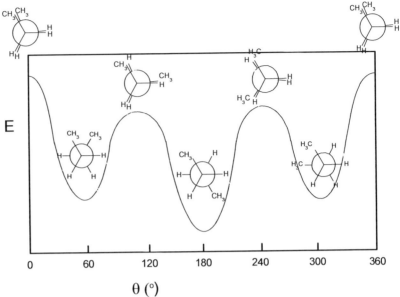

3. Conformations of butane:

gauche   anti   gauche

4. The following saw-horse projection represents the gauche conformation of butane indicated by a formula on the previous page:

5. Main factors that determine stability of compounds such as 1-bromo-2-chloroethane are individual bond dipoles rather than steric interactions. In general, in the lowest energy conformation individual bond dipoles orient themselves to compensate for each other (point in opposite directions). If bulky substituents are also present then the lowest energy conformation is a result of interplay between steric and electronic factors and the analysis may be rather complex. In case of 1-bromo-2-chloroethane, we need to consider only the individual bond dipoles.

anti (the most stable)   or   syn (the least stable)

6. This is a difficult question and only a very few students are able to answer it correctly. The main stabilizing interaction in 1,2-ethanediol is an intamolecular hydrogen bond. Recall that hydrogen bond is the strongest non-bonding interaction. In an eclipsing (syn) conformation the two –OH groups are in the best position to form an intramolecular hydrogen bond. The actual conformation is somewhere between a Gauche and a syn conformation. In a real Gauche conformation oxygen and hydrogen atoms that form hydrogen bond are too far apart and are not properly aligned (recall that a hydrogen bond is directional unlike other intermolecular interactions). In a syn conformation there is still torsional stain. Thus, a molecule twists a bit to minimize torsional strain while still allowing hydrogen bonding. The Newman projection on the left shows such conformation.

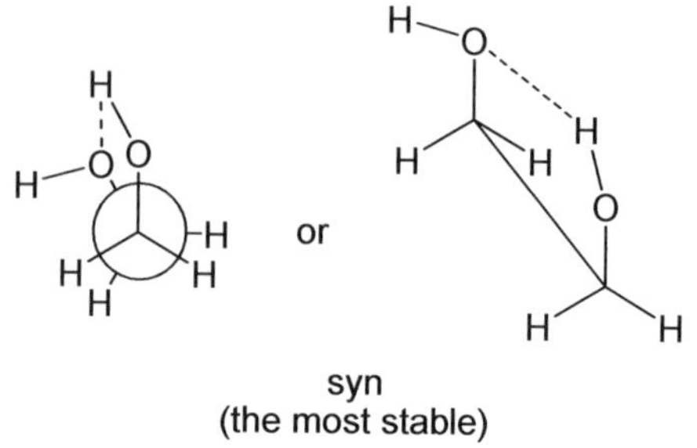

syn
(the most stable)

Questions 7-15.

7. Draw a chair and a boat conformation of cyclohexane and indicate types of strain present in each conformation (if any).

8. Draw all the possible conformations of 1,3-dimethylcyclohexanes. Identify the most stable structure.

9. Draw the possible conformations of *cis*-1-*t*-butyl-3-methylcyclohexane. Identify the most stable structure.

10. Draw a Newman projection of **the most stable** conformation of butane. What types of strain are present in the conformation (if any)?

11. Draw a chair conformation of cyclohexane. Include in the drawing all of the axial and equatorial substituents (label each substituent as either axial or equatorial).

12. Which of the Newman projections shown below represents the conformation shown by the saw-horse projection?

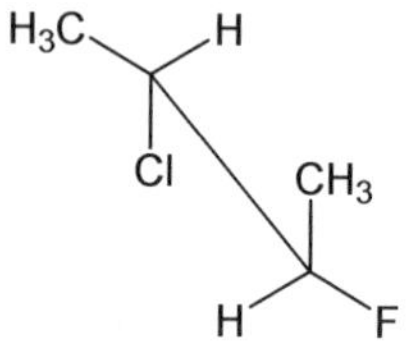

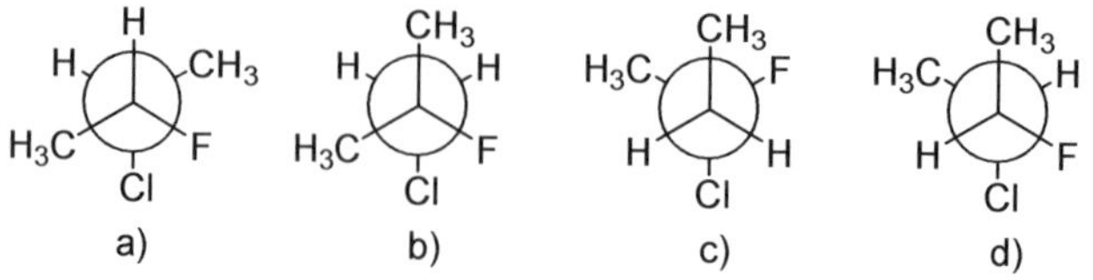

a)   b)   c)   d)

13. Which of the following statements is correct?

   a) Cyclopropane is more stable than cyclopentane due to cyclopentane's angle strain.
   b) Cyclopropane is more stable than cyclopentane due to cyclopropane's angle strain.
   c) Cyclopropane is less stable than cyclopentane due to cyclopentane's angle strain.
   d) Cyclopropane is less stable than cyclopentane due to cyclopropane's angle strain.

14. Cyclobutane has the following type(s) of strain:

a) torsional
b) torsional and steric
c) torsional and angle
d) steric and angle

15. Gauche conformation of butane has the following type(s) of strain:

a) torsional
b) angle
c) steric
d) torsional and steric

Answers to questions 7-15.

7. Chair and boat conformations of cyclohexane:

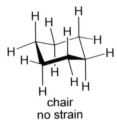

chair
no strain

boat
large torsional strain
small steric strain

8. There are two isomers of 1,3-dimethylcyclohexane – *cis* and *trans*. Each has two conformations. In each case, the conformation with a larger number of equatorial substituents is the more stable one.

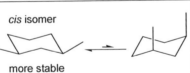

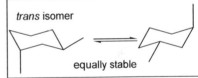

9. In each of the two possible chair conformations of *cis*-1-*t*-butyl-3-methylcyclohexane one substituent is axial and one equatorial. A conformation in which the larger substituent is equatorial is the more stable one.

more stable
(*t*-butyl equatorial)

10. The most stable conformation of butane:

$$\text{anti (no strain)}$$

11. Chair conformation of cyclohexane with labeled substituents (a-axial, e-equatorial):

12. d); 13 d); 14. c); 15. c).

Exercises in Organic Chemistry

Test 1: CHAPTERS 1-4

1. Use the valence Bond Theory to describe the bonding in ethyne (acetylene). A complete answer will include a Lewis structure of the molecule, the electronic configuration of carbon showing hybridization, and a sketch of the molecule showing the types of bonds formed.

2. Calculate formal charges of the labeled atoms:

3. Draw the complete structural formula (including all the bonds and all the hydrogens) for the following compound:

4. Classify the following compounds as primary, secondary, tertiary or quaternary:

5. The following are incorrect names for organic molecules. Give correct name for each compound.

   a) 3-methyl-4-propylpentane
   b) 2-ethyl-3-methyl-5-ethylhexane
   c) 2-ethyl-4-methylpentane

6. Draw the structural formula (including all the hydrogens) for each of the following compounds:

   a) 3-methylnonane
   b) 1-bromo-3-ethylpentane

7. Complete the following table giving the structure or name for each molecule.

Name	Structural formula
	(structure with NH$_2$)
3-chloro-1-pentanol	
2-phenylethylamine	
	(structure with Cl, phenyl, and OH)
ethylphenylamine	
	(structure with NH$_2$ and OH)
aniline	

8. The C-O-H bond angle in ethanol is approximately:
   a) 90°
   b) 109°
   c) 120°
   d) 180°

9. Draw and name Newman projections of the *most stable* and the *least stable* conformations of 1,2-dichloroethane.

Exercises in Organic Chemistry

10. Draw the *most stable* conformations of the following compounds:

    a) *trans*-1,2-dimethylcyclohexane
    b) *trans*-1,3-dichlorocyclohexane
    c) *cis*-1,4-dimethylcyclohexane

11. Heterolytic cleavage of the C-Y bond in the compound CH_3Y gives:

    a) two free radicals
    b) a carbocation ion
    c) a carbanion
    d) either a carbocation or a carbanion

Test 1 – Solutions

1. The bonding in ethyne: H—C≡C—H

    C $1s^2\ 2s^2\ 2p^2$

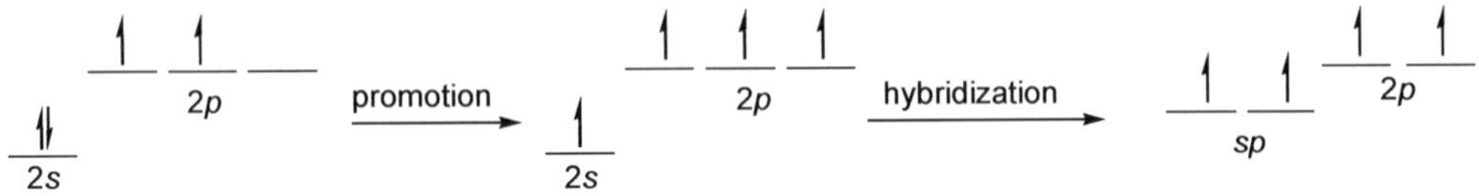

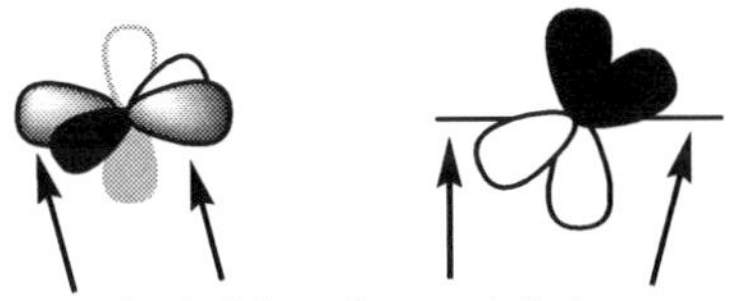

    hybridized *sp* orbitals

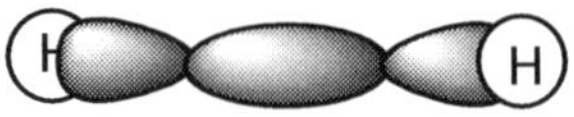

    σ bonds in ethyne

    σ and π bonds in ethyne

2. Formal charges of the labeled atoms:

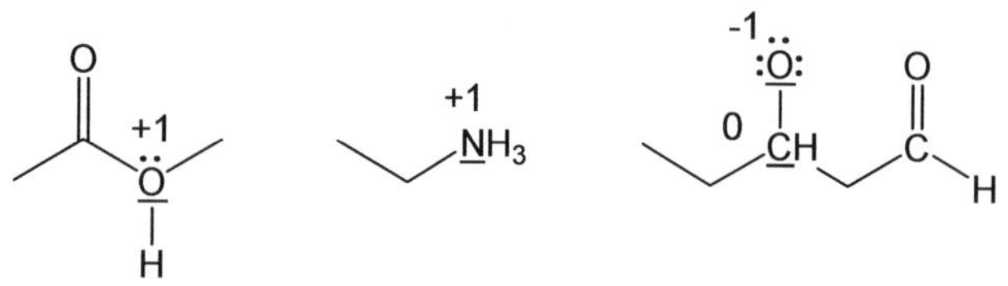

3. The complete structural formula:

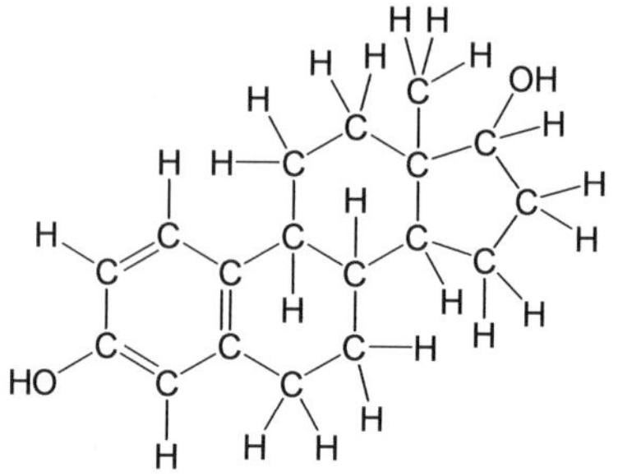

4. Degree of substitution:

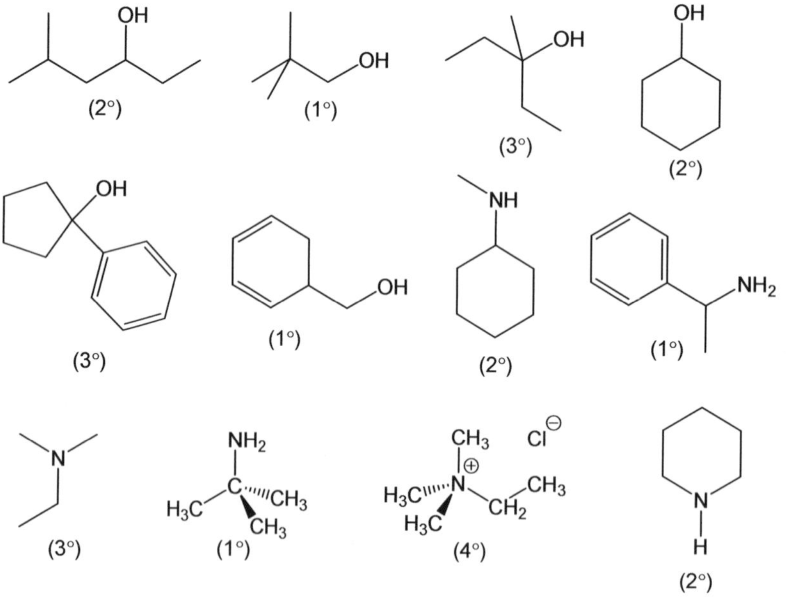

5. Correct names are: a) 3,4-dimethylheptane; b) 3,4,6-trimethyloctane; c) 2,4-dimethylhexane.

6. Structural formulas:
   a) 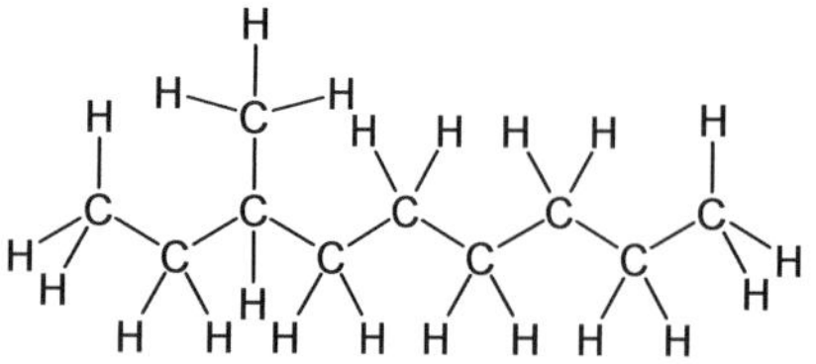

   b)

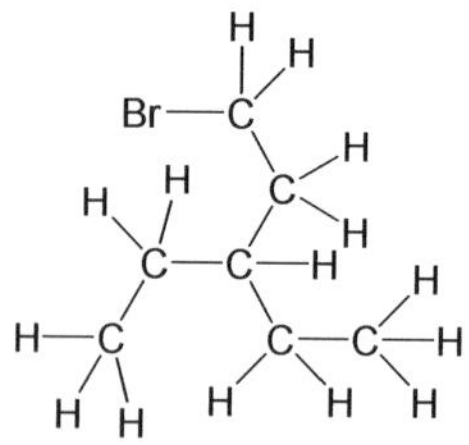

7. Completed table:

Name	Structural formula
3-hexanamine	(structure with NH₂)
3-chloro-1-pentanol	(structure with Cl and OH)
2-phenylethylamine	(structure with H₂N and phenyl)
4-chloro-4-phenyl-1-hexanol	(structure with Cl, phenyl, and OH)
ethylphenylamine	(structure with HN and phenyl)
5-amino-1-hexanol	(structure with NH₂ and OH)
aniline	(phenyl with NH₂)

8. c).

9. Newman projections of the *most stable* and the *least stable* conformations of 1,2-dichloroethane.

most stable (anti)   least stable (Cl's are eclipsed)

10. The *most stable* conformations:
   a)
   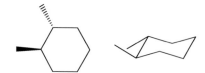

   b)

   c)

11. d);

Exercises in Organic Chemistry

5. Acids and Bases

Concepts to review: Brønsted-Lowry and Lewis definitions of acids and bases, pK_a and pH, acid-base equilibrium, organic acids and bases, effect of structure on acid strength, effect of pH on structure of organic compound.

Questions 1-2.

1. The following are pK_a values, given in no particular order, of the acids below. Assign each pK_a value to the corresponding acid.

    pK_a values: 4.74; 0.64; 2.86; and 1.26

    a) CH_3CO_2H _____
    b) $ClCH_2CO_2H$ _____
    c) CCl_3CO_2H _____
    d) Cl_2CHCO_2H _____

2. Write the **Lewis formulas** of sodium hydroxide, nitric acid, phenol, ethyl alcohol, and acetic acid.
    a) Are there any common structural elements for all of these compounds?
    b) What determines whether a compound is acidic, basic or neutral?
    c) Arrange these compounds in order of increasing acidity.

3. Arrange the following acids in the order of *increasing* acid strength.

    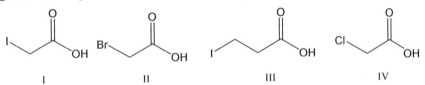

    I II III IV

    a) II < I < IV < III
    b) IV < II < I < III
    c) III < IV < I < II
    d) III < I < II < IV

4. Arrange the following acids in the order of increasing strength.

35

Answers to questions 1-4.

1. To answer this question correctly, one has to know the factors that affect strength of carboxylic acids and be able to apply that knowledge. In addition, one has to know that a stronger acid has a lower pK_a value compared to a weaker acid. Therefore, the order of pK_a values from the strongest acid to the weakest is: 0.64, 1.26, 2.86 and 4.74.

All of the four carboxylic acids have two carbon atoms. Thus, they are acetic acid and chlorine substituted derivatives of acetic acid. Due to their inductive effect, chlorine atoms increase acid strength, which means that chloroacetic acid is a stronger acid than acetic acid. Inductive effects are additive. Therefore, dichloroacetic acid is stronger than chloroacetic acid, and trichloroacetic acid is even stronger than dichloroacetic acid. Order of acid strength from the weakest to the strongest is:

CH_3CO_2H	$ClCH_2CO_2H$	Cl_2CHCO_2H	CCl_3CO_2H
acetic acid	monochloroacetic acid	dichloroacetic acid	trichloroacetic acid

and their pK_a values are:

a) CH_3CO_2H 4.74
b) $ClCH_2CO_2H$ 2.86
c) CCl_3CO_2H 0.64
d) Cl_2CHCO_2H 1.26

2. Structural formulas are:

- Na⁺ ⁻O—H   sodium hydroxide
- H—O—N⁺(=O)—O⁻   nitric acid
- C₆H₅—O—H   phenol
- H_3C-CH_2-O-H ethanol
- $H_3C-C(=O)-O-H$ acetic acid

a) All of the above compounds have an –OH group and can be represented with a general formula Y-O-H.
b) Acidity of a compound depends on the relative strengths of Y-O and O-H bonds. Compounds with weak highly polarized O-H bond are strong acids. Heterolytic breaking of O-H bond gives H^+ ions. On the other hand, compounds with weak highly polarized Y-O bond are basic rather than acidic – they are sources of OH⁻ ions.
c) Order of increasing acidity is:

Na⁺ ⁻O—H < H_3C-CH_2-OH < C₆H₅OH (phenol) < $H_3C-C(=O)-OH$ < HO—N⁺(=O)—O⁻

weakest acid   strongest acid

3. .A question of this type should be treated as an open ended question. Therefore, it is the best to answer it on your own and then find which one of the four matches your answer. Halogen substituents increase acid strength due to their negative inductive effect, which in turn is a consequence of their electronegativity. Thus, given the same distance from the –OH group, the most electronegative substituent (-Cl) increases acid strength the most and least electronegative (-I) the least. Finally, iodine in the position 3 has less of an effect than iodine in the position 2 and the order of increasing acid strength is: III < I < II < IV, or d).

4. This question is considerably more complex compared to the questions 1 and 3. In addition to inductive effects, one has to consider resonance effects and know their relative importance. By now, it should be easy to arrange acetic and chloroacetic acids in the order of increasing acid strength based on the inductive effect of chlorine substituent: III<I<II. Ethanol (IV) has no inductive and resonance effects of a carbonyl group (C=O) to increase its acidity and is, therefore, the weakest acid. Now the order of acidity is; IV< III<I<II. Finally, -OH group of phenol (V) experiences a negative resonance effect of benzene ring. Effect of benzene ring on acidity is important, but not as strong as that of a carbonyl group. Therefore, phenol is weaker acid than acetic acid and stronger than ethanol and the overall order is:

$$IV<V<III<I<II$$

Questions 5-7.

5. Arrange the following acids in the order of increasing strength.

6. Arrange the following acids in the order of increasing strength.

7. Arrange the following acids in the order of increasing strength.

Answers to questions 5-7.

5. Do not be confused by a ring! Treat these acids the same way you would treat acyclic (open-chain) acids. Therefore, you should consider inductive effect of fluorine atom and the decrease of it with distance. Order of acid strength from the weakest to the strongest is:

$$IV<II<I<III$$

6. All of the six carboxylic acids have four carbon atoms - they are: trifluoroacetic acid **VI**, butanoic acid **II** and fluorine substituted derivatives of butanoic acid **I**, **III**, **IV** and **V**. As in the question 1 one should consider the inductive effect, this time of a fluorine atom. Fluorine atoms are even more electronegative than chlorine and, as substituents, increase strength of an acid. Therefore, the acid without any fluorine substituent, butanoic acid (**II**) is the weakest acid. Inductive effects are additive and decrease with increasing distance from the –OH group. It means that 2,2,2-trifluoroacetic acid (**VI**) is the strongest in the group as it has the largest number of fluorines and they are as close as possible to the –OH group. Next strongest is 2,2-difluorobutanoic acid (**V**), followed by 2-fluorobutanoic acid (**IV**). Strength of the remaining two acids decreases with increasing distance between fluorine and –OH. Thus, 4-fluorobutanoic acid (**III**) is weaker than 3-fluorobutanoic acid (**I**). Overall order of acid strength from the weakest to the strongest is:

$$II<III<I<IV<V<VI$$

7. To answer this question, one should keep in mind both that the inductive effects are additive as well as that they increase in the order of I < Br < Cl < F. Thus, it should be easy to identify the weakest and the strongest acid. The acid with the two bromine substituents, dibromoacetic acid **II**, is the weakest acid while the acid with the two fluorine substituents, difluoroacetic acid **III**, is the strongest. Both of the remaining two acids have a chlorine atom as one of the substitiuents and a difference in acid strength is due to the remaining substituent. In case of bromochloroacetic acid (**I**) that substituent is bromine, while in the case of chlorofluoroacetic acid (**IV**) it is fluorine. Therefore, acid **IV** is stronger than acid **I** and the overall order of acid strength is:

$$II < I < IV < III$$

Additional Exercises: Questions 8-16.

8. Arrange the following acids in the order of increasing strength.

9. Arrange the following acids in the order of increasing strength.

Exercises in Organic Chemistry

10. Which of the following carboxylic acids has the largest pK_a value?

    a) CH_3CO_2H
    b) $ClCH_2CO_2H$
    c) CF_3CO_2H
    d) $CH_3(CH_2)_2CO_2H$

11. The anion of a weak acid is:

    a) a strong base.
    b) a weak base.
    c) an electrophile.
    d) electron deficient.

12. Which one of the following compounds is the strongest acid?

    $HC\equiv C-CH_3$ $H_3C-C\equiv C-CH_3$ $H_3C-\underset{H}{\overset{CH_3}{\underset{|}{C}}}-\overset{}{\underset{|}{C}}-CH_3$ ⬠

    a)   b)   c)   d)

13. Resonance effect is:

    a) withdrawal of electrons along a π-bond
    b) withdrawal of electrons from an –OH group
    c) displacement of electrons along a π-bond
    d) displacement of electrons along a σ-bond

14. In the following examples acids are arranged in the order of decreasing acid strength. Which of the following series is incorrect?

    a) $CH_3CH_2CHFCOOH > CH_3CHFCH_2COOH > CH_3CH_2CH_2COOH$
    b) $CH_3CHBrCOOH > CH_3CHClCOOH > ClCH_2CH_2COOH$
    c) $Br_3CCOOH > BrCH_2COOH > ICH_2COOH$
    d) $CH_3COOH > CH_3CH_2COOH > CH_3CH_2CH_2COOH$

15. Aniline, $C_6H_5NH_2$, has a $pK_b = 9.4$. The pK_a of its conjugate acid, $C_6H_5NH_3^+$, is:

    a) 4.6
    b) 4.7
    c) 18.8
    d) 0.6

16. Which acid is most highly dissociated in an aqueous solution?

    a) CH_3CBr_2COOH
    b) Br_2CHCH_2COOH
    c) Cl_2CHCH_2COOH
    d) CH_3CCl_2COOHd

39

Answers to questions 8-16.

8. II<I<IV<III
9. I<II<III<IV
10. d); 11. a); 12. a); 13. c); 14. b); 15. a); 16. d).

Exercises in Organic Chemistry

6. Organic Reactions

Concepts to review: electrophiles and nucleophiles, Lewis acids and bases, thermodynamics and kinetics of a reaction, The Hammond postulate, Principle of Microscopic Reversibility, pushing electrons.

Questions 1-5.

1. In each of the following five pairs circle the stronger nucleophile

Cl$^-$	or	HCl
HO$^-$	or	HS$^-$
Br$^-$	or	I$^-$
H$_3$P	or	H$_2$S
Br$^-$ (in water)	or	Br$^-$ (in acetone, an aprotic solvent)

2. The order of increasing strength of the following nucleophiles is:

   $^\ominus$NH$_2$ $^\ominus$I $^\ominus$F $^\ominus$OH
     I   II   III   IV

   a) II < I < IV < III
   b) II < IV < I < III
   c) III < IV < I < II
   d) III < I < IV < II

3. Which of the following is an electrophile?

   a) NH$_4^+$
   b) CH$_3$OH
   c) AlCl$_4^-$
   d) BCl$_3$

4. Which of the following represents a correct electron flow?

   a)   b)   c)   d)

5. Consider the following reaction diagram.

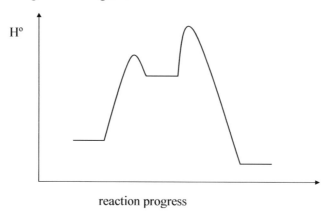

reaction progress

a) Label the reactants and the products.
b) Label the intermediates and the transition states.
c) Is the overall reaction exothermic or endothermic?
d) How many steps there are in the reaction mechanism?
e) Which step is the rate determining step?

Answers to questions 1-5.

1. When considering strength of nucleophiles one should take into account the following factors
(a) If the elements with free electron pairs belong to the same row of the periodic table, then the stronger base is the stronger nucleophile. Usually, that is the element to the left.
(b) If we are comparing nucleophiles where the free electrons are on the same element, then one with a negative charge is the stronger nucleophile.
(c) If the elements with electron pair belong to the same group of the periodic table, then more polarizable (larger) element is a stronger nucleophile.
(d) Steric effects may reduce strength of nucleophiles.
(e) Polar protic solvents exhibit a "leveling effect" on strength of nucleophiles due to formation of either hydrogen bonds or strong dipole-dipole interactions.

In each of the five pairs the stronger nucleophile is:

Cl⁻	or	HCl	Cl⁻ point (b).
HO⁻	or	HS⁻	HS⁻ point (c).
Br⁻	or	I⁻	I⁻ both points (b) and (c).
H₃P	or	H₂S	H₃P point (a).
Br⁻ (in water)	or	Br⁻ (in acetone, an aprotic solvent)	Br⁻ (in acetone) point (e).

2. The best is to treat this question as an open one. Applying the point (a) above we get the order of strength of three of the nucleophiles: F⁻<HO⁻<NH₂⁻. Finally, as by far the most polarizable ion (point (c)), iodide is the strongest nucleophile. Therefore, the overall order is F⁻<HO⁻<NH₂⁻<I⁻ or III<IV<I<II and the correct answer is c).

3. By definition an electrophile has an empty orbital. This question should be treated as a closed question, which means that we eliminate wrong answers until we are left with the correct one. a) NH_4^+ has the octet on the nitrogen filled and is not an electrophile. This one often confuses students because of the positive charge on nitrogen. A positive charge does not mean that a species is an electrophile. It is the empty orbital that counts. b) methanol has both oxygen and carbon octets filled and is not an electrophile either. c) $AlCl_4^-$ has no empty orbitals either and is actually a source of hydride ions – nucleophiles. This leaves us with d) BCl_3 as the electrophile. Boron is electron deficient and has an empty orbital.

4. A complete curved arrow indicates movement of two electrons, while a half-arrow ("fish-hook arrow") indicates movement of a single electron. Complete arrows always follow each other in head to tail fashion. Therefore, a) is wrong because they point head to head. It is fish-hook arrows that always point head to head or tail to tail. Therefore, d) where they follow each other is wrong. Answer b) has a full arrow and fish-hook arrow. That is not possible since a mechanism either involves free radicals (fish-hook arrows) or nucleophile/electrophile interaction (full arrows), but not a mix of the two. This leaves us with c), where two fish-hook arrows point towards each other, as the correct answer.

5. The overall reaction is exothermic – the energy of the products is lower than that of the reactants. The reaction has two steps (two transition states). The second step is the rate-determining step. Transition state that corresponds to the second step is the overall maximum on the energy plot.

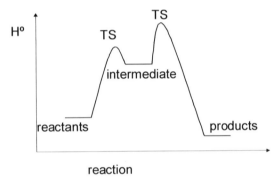

Additional Exercises: Questions 6-14

6. Consider the following reaction diagram.

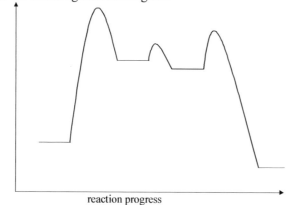

   a) Label the reactants and the products.
   b) Is the overall reaction exothermic or endothermic?
   c) Label the intermediates and the transition states.
   d) How many steps there are in the reaction mechanism?
   e) Which step is the rate determining step?

7. Which of the following is not a nucleophile?

   a) CH_3OH
   b) H_2
   c) CN^-
   d) NH_3

8. Which of the following series contains a free radical, a nucleophile and an electrophile in that order?

   a) Cl_2, NH_4^+, NH_3
   b) $\cdot CH_3$, NH_3, OH^-
   c) $\cdot Br$, H_2O, BF_3
   d) CH_4, OH^-, CO_2

9. Series: H^+, OH^-, $H:^-$, H_2O contains:

   a) only nucleophiles
   b) only electrophles
   c) one electrophile and three nucleophiles
   d) two nucleophiles and two electrophiles

10. K⁺ is not an electrophile because:

    a) it does not form a covalent bond with a nucleophile.
    b) it does not have an empty orbital.
    c) it is not electron deficient.
    d) it has its octet filled.

11. Which of the following series contains only electrophiles:

    a) NH_3, H_2O, Cl^-, I^-
    b) $AlCl_3$, NH_3, H_2O, Br^-
    c) $AlCl_3$, BF_3, H^+, Br_2
    d) $AlCl_3$, BF_3, NO_2^+, I^-

12. Hydride ion is:

    a) an electrophile
    b) a nucleophile
    c) a Brønsted-Lowry acid
    d) a Lewis acid

13. Which compound has a carbon-hydrogen bond with the lowest bond dissociation energy?

    a) ethene
    b) benzene
    c) ethyne
    d) ethane

14. A forward reaction has three steps with the first step being the rate determining step. The reverse reaction:

    a) has three steps with the last step being the rate determining.
    b) has three steps with the first step being rate determining.
    c) has two steps with the last step being the rate determining.
    d) has two steps with the first step being the rate determining.

Answers to questions 6-14

6. The overall reaction is exothermic, has three steps and the first step is the rate determining step.

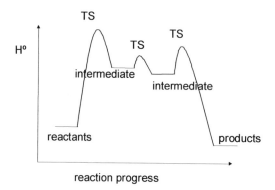

7. b); 8. c); 9. c); 10. a); 11. c); 12. b); 13. d); 14. a).

Exercises in Organic Chemistry

7. Reactions of Alkanes

In an introductory organic chemistry course we study only free radical reactions of alkanes. Other types of reactions such as fragmentation and combustion are not studied. When it comes to free radical reactions of alkanes, there are four types of questions one can encounter:

1) Provide formulas of all possible substitution products.
2) Given the yields of different products, calculate intrinsic reactivities.
3) Given the intrinsic reactivity, calculate the expected yields.
4) Questions related to mechanism of free radical chain reaction.

Questions 1-4.

1. Provide structural formula of all possible products of the following reaction. Assume that only monosubstitution products are obtained.

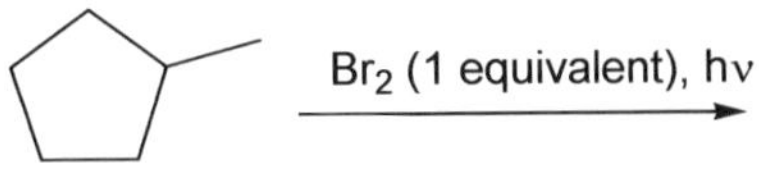

    cyclopentane + Br_2 (1 equivalent), $h\nu$ →

2. Calculate the intrinsic reactivity towards free radical bromination of a tertiary carbon atom compared to the reactivity of a secondary carbon atom from the data given below.

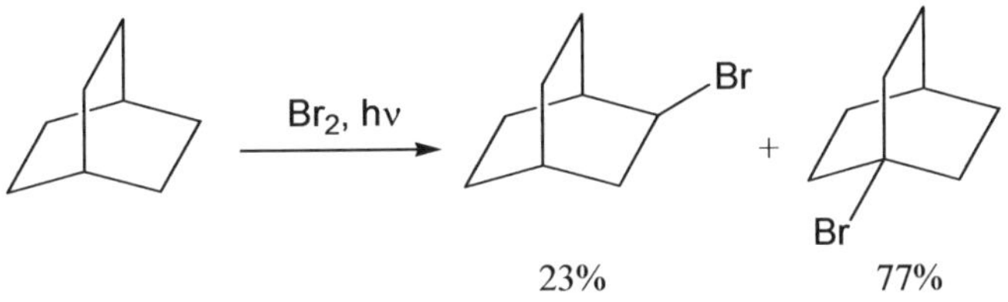

    bicyclic alkane + Br_2, $h\nu$ → product with Br (23%) + product with Br (77%)

3. The intrinsic reactivity towards free radical chlorination of secondary carbon atom is 4 times higher compared to the intrinsic reactivity of a primary carbon atom. Calculate either the expected ratio, or the expected yield of the two monochlorination products.

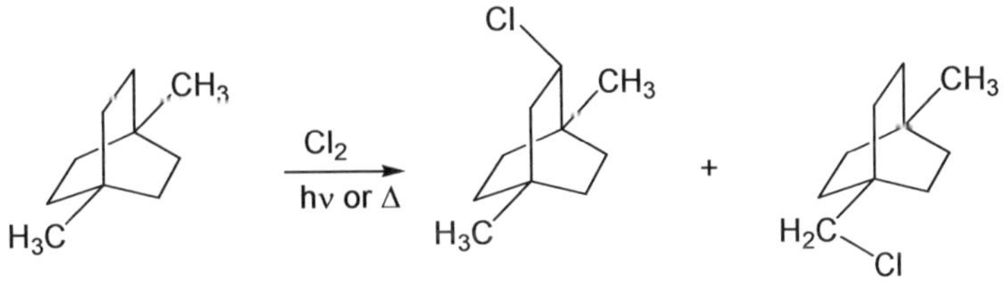

    substrate + Cl_2, $h\nu$ or Δ → two products shown

4. Which of the following correctly represents electron flow for the bromination of an alkene?

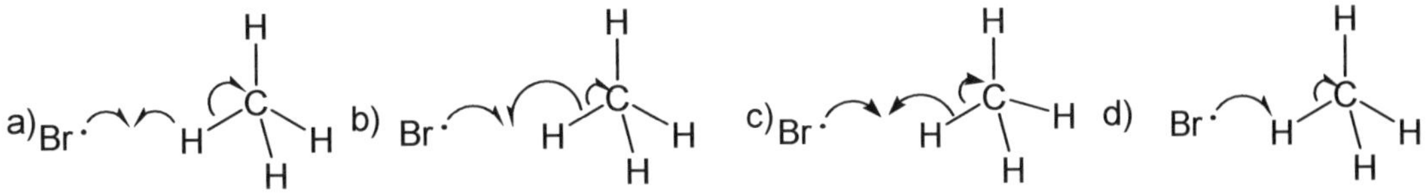

    a) Br· ... b) Br· ... c) Br· ... d) Br· ...

Answers to questions 1-4.

1. Monosubstitution means that in each molecule only one of the hydrogens is replaced by a chlorine atom. To find a number of monosubstitution products, one has to find how many different carbons are in the molecule and then place a chlorine atom on each one. Thus, methylcyclopentane has four different carbons. They are shown here labeled with different letters: 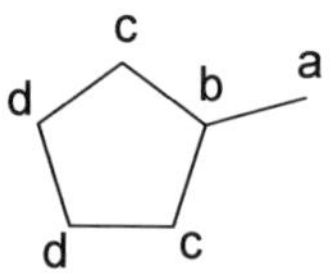
and the four reaction products are:

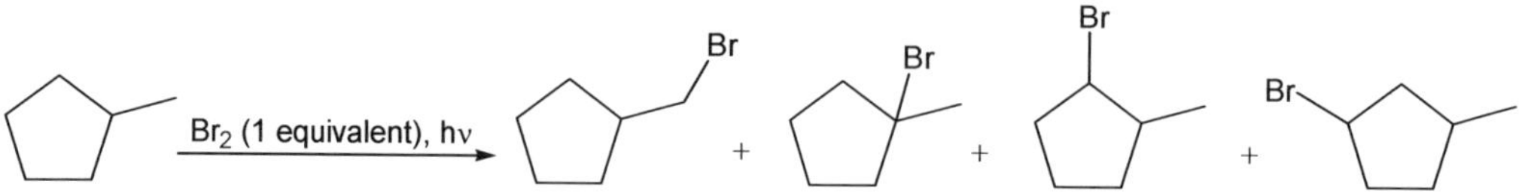

2. There are two hydrogens on the tertiary carbon atoms and twelve on the secondary. Substitution of the two hydrogens on tertiary carbons gives 77% of the product and substitution of the twelve on the secondary gives 23% of the product. The intrinsic reactivity tertiary/secondary is (77/2)/(23/12) = 20. Therefore, a tertiary carbon is 20 times more reactive than secondary towards bromination.

3. The expected ratio is (number of secondary x intrinsic ratio)/(number of primary x 1). There are 12 hydrogens on secondary carbons and 6 on the primary. Therefore, the calculation is:
    (12 x 4)/(6 x 1) = 8, which means that the ratio of the two products is 8:1, or expressed as percent yield 89%:11%.

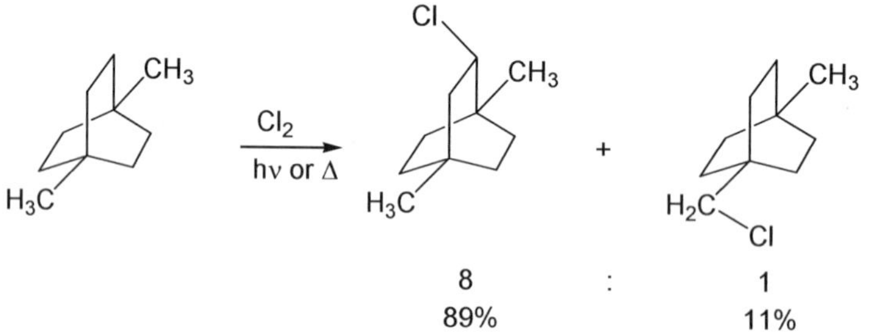

4. Students frequently find this question to be difficult. One has to be very careful in considering details of the mechanism and the electron movement. Answer a) has one half arrow pointing from the hydrogen atom towards the bromine's free electron, which is incorrect. Answer b) is the correct one. Half arrows show homolytic breaking of the carbon-hydrogen bond and formation of hydrogen-bromine bond. Answer c) contains full arrows, which indicate movement of electron pairs instead of single electrons, instead of half arrows. That is obviously incorrect. However, many students miss that and do not notice any difference between the answers b) and c). Answer d) shows two half arrows following one another, which is incorrect.

Exercises in Organic Chemistry

Additional Exercises: Questions 5-14

5. The intrinsic reactivity towards free radical bromination of secondary carbon atom is 72 times higher compared to the intrinsic reactivity of a primary carbon atom. Calculate either the expected ratio, or the expected yield of the monobromination products.

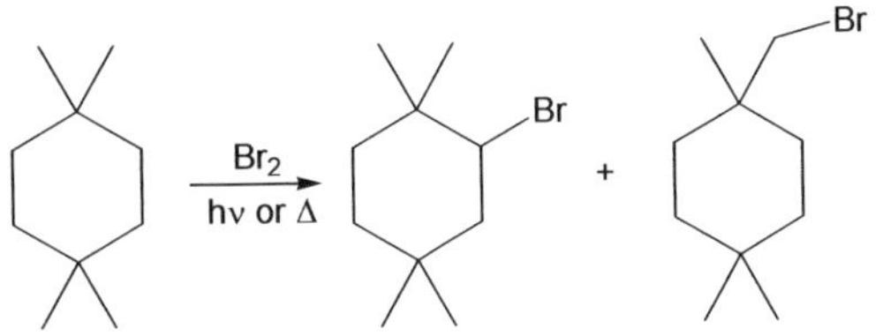

6. The intrinsic reactivity towards free radical chlorination of secondary carbon atom is 4 times higher compared to the intrinsic reactivity of a primary carbon atom. Calculate either the expected ratio, or the expected yield of the monochlorination products.

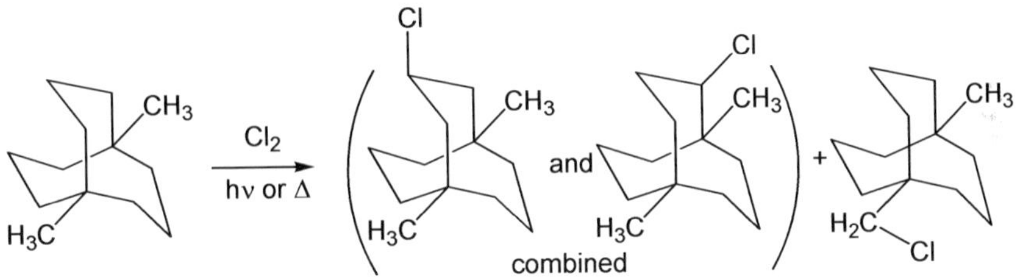

7. Free radical chlorination of adamantane gives a mixture of two products: 1-chloroadamantane and 2-chloroadamantane. The intrinsic reactivity towards free radical chlorination of a tertiary carbon atom is 1.3 times higher compared to the intrinsic reactivity of a secondary carbon atom. Calculate either the expected ratio, or the expected yield of the monochlorination products.

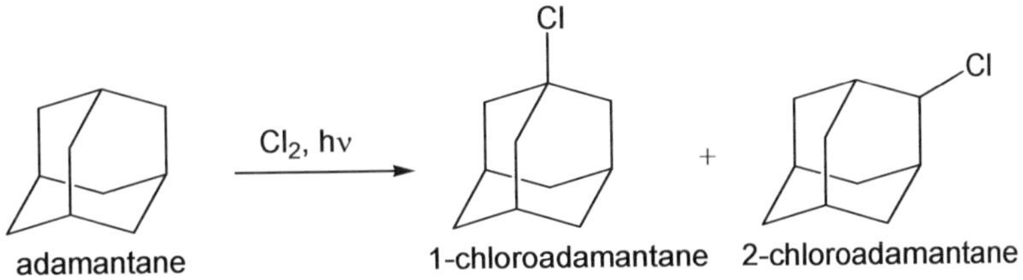

8. Provide structural formula of all possible products of the following reaction. Assume that only monosubstitution products are obtained.

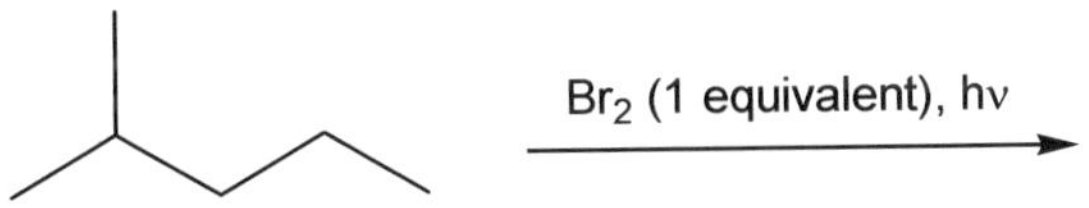

9. Provide structural formula of the major product of the following reaction.

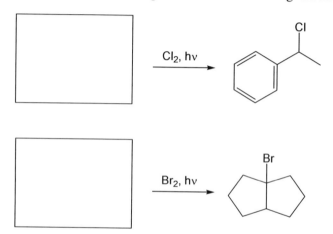

10. Provide structural formulas of the starting materials of the following reactions.

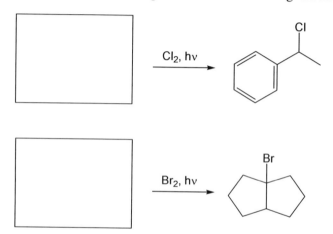

11. An alkane is most likely to react with:

   a) an electrophile.
   b) a nucleophile.
   c) a Lewis acid.
   d) a free radical.

12. In a radical chlorination of methane, which of the following represents a propagation step?

   a) Cl· + CH₄ → ·CH₃ + HCl
   b) HCl + heat → H· + ·Cl
   c) ·CH₃ + ·CH₃ → CH₃-CH₃
   d) ·CH₃ + ·Cl → CH₃Cl

13. The free radical reaction below might produce the following products. Which compound is the major or the predominant product?

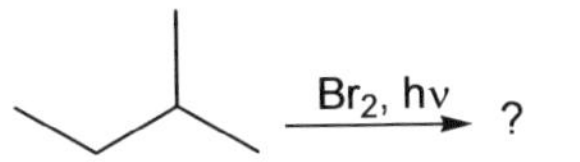

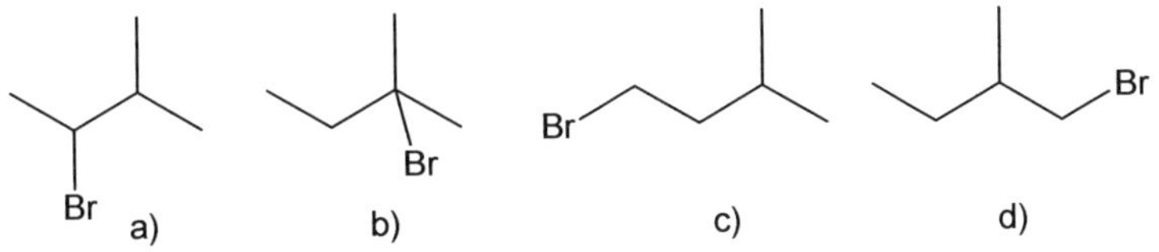

14. How many products can be formed in a free radical bromination of methane?

   a) 1
   b) 2
   c) 3
   d) 4

Answers to Additional Exercises: Questions 5-14

5. The expected ratio and the expected yield of the monobromination products:

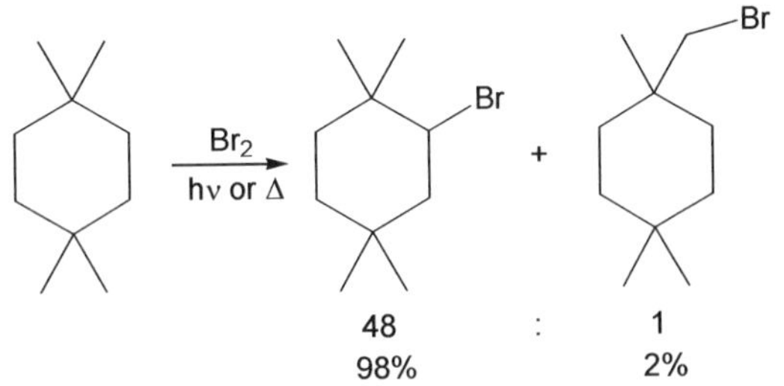

   48 : 1
   98% 2%

6. The expected ratio and the expected yield of the monochlorination products.

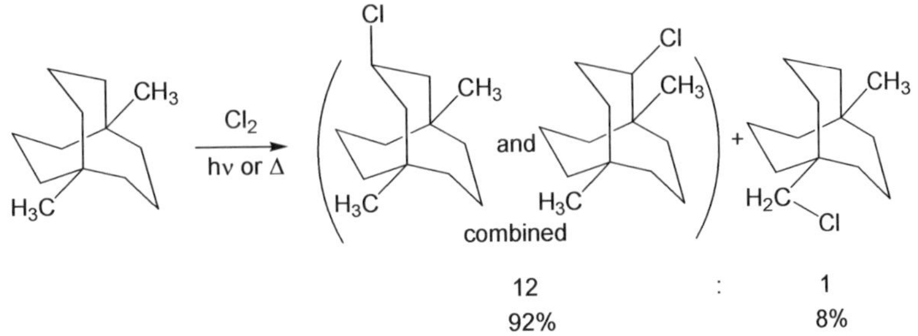

   12 : 1
   92% 8%

51

7. The expected ratio and the expected yield of the monochlorination products:

adamantane → (Cl₂, hv) → 1-chloroadamantane + 2-chloroadamantane

3.9 : 1
80%   20%

8. There are five products:

[Five isomeric bromide products shown]

9. Reaction product:

cyclohexene + N-bromosuccinimide → (hv) → 3-bromocyclohexene

10. The starting materials:

ethylbenzene → (Cl₂, hv) → (1-chloroethyl)benzene

bicyclo[3.3.0] system → (Br₂, hv) → bridgehead bromide

11. d); 12. a); 13. b); 14. d).

Exercises in Organic Chemistry

8. Alkenes and Alkynes: Nomenclature, Structure and Physical Properties

Concepts to review: degree of unsaturation, nomenclature and isomerism of alkenes and alkynes, Cahn-Ingold-Prelog sequence rules, structure and physical properties of alkenes and alkynes.

Questions 1-5.

1. Assign correct IUPAC names to the following compounds.

2. An unsaturated compound with one triple bond and no double bonds has a molecular formula $C_{10}H_{16}$. How many rings does the molecule have?

   a) 0
   b) 1
   c) 2
   d) 3

3. A compound C_5H_7Br has:

   a) Two double bonds
   b) Two double bonds and a ring
   c) Two triple bonds
   d) Two double bonds and a triple bond

4. The following are the boiling points of three unsaturated hydrocarbons. Explain the trend.

   1°C   4°C   27°C

5. Convert the multiple bonds in the following two substituents into single bonds by adding virtual atoms. Which of the two substituents has the higher priority? What is the Canh-Ingold-Prelog designation of the stereochemistry of the double bond?

Answers to questions 1-5.

1. IUPAC names:

   1-ethylcyclopentene
   (note that ethyl must be numbered)

   3-propyl-2-hexene

   (E)-2-pentene

   3-vinylcyclopentene

   1-allylcyclopentene

   3-butyl-1-hexene

2. First one should assess the question. Is it closed, or open ended? This one is open ended. The question itself contains enough information so that it can be answered without considering the four possible answers. From the molecular formula one can calculate the Degree of Unsaturation (also called Index of Hydrogen Deficiency) of the molecule. Number of Degrees of Unsaturation equals the sum of π bonds and rings in a molecule.

$$DoU = [(2 \times n + 2) - m]/2 = [(2 \times 10 + 2) - 16]/2 = 3,$$

where n is the number of carbon atoms and m is the number of hydrogen and halogen atoms

Therefore, the molecule has three degrees of unsaturation. Since we know that there are no double bonds and that there is one triple bond (2 π bonds), which accounts for two degrees of unsaturation, the remaining unsaturation must be a ring. Thus, there is one ring in a molecule and the correct answer is b).

3. This question is similar to the previous one. A difference is that this one is a closed question. Therefore, one has to consider the four possible answers and to choose the right one. Formula C_5H_7Br indicates that the compound has two degrees of unsaturation.

$$DoU = [(2 \times n + 2) - m]/2 = [(2 \times 5 + 2) - (7+1)]/2 = 2$$

Next, one should eliminate all the wrong answers. Answer b) has 3 degrees of unsaturation, answers c) and d) have four degrees of unsaturation each. Only answer a) has two degrees of unsaturation and is the correct answer.

4. Boiling points are a reflection of the strength of intermolecular interactions. *Trans*-2-butene exhibits only relatively weak van der Waals interactions and has the lowest boiling point, *cis*-2-butene, in addition to the approximately same magnitude van der Waals interactions, exhibits a weak dipole-dipole interaction (recall that *cis* alkenes are polar, while symmetrical *trans* alkenes are not), which results in slightly higher boiling point. Finally, 2-butyne exhibits relatively strong van der Waals interactions due to the presence of highly polarizable triple bond.

5. The two substituents are a methyl ester and amide groups:

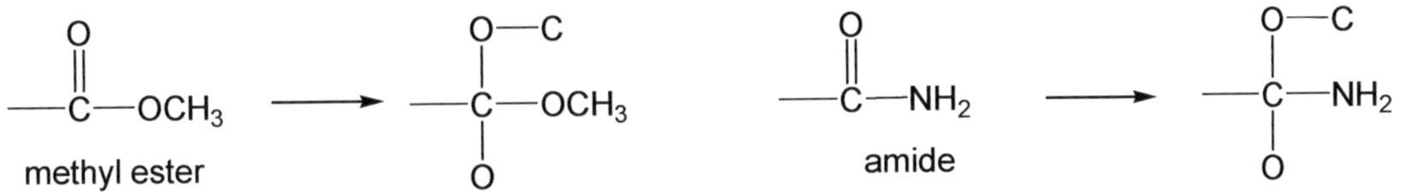

Exercises in Organic Chemistry

methyl ester: —C(=O)—OCH₃ → —C(O—C)(OCH₃)—O

amide: —C(=O)—NH₂ → —C(O—C)(NH₂)—O

Since oxygen has a higher atomic number than nitrogen, it has a higher priority and, in turn, methyl ester substituent has a higher priority compared to the amide substituent. Therefore, the alkene is (*E*)- isomer.

Additional Exercises: Questions 6-10

6. Complete the following table by providing the structural formulas of the following compounds.

3-methylcyclopentene	
(*Z*)-4-bromo-2-pentene	
3,3-dimethylcyclopentene	
4-bromo-2-pentyne	
1-bromo-1-pentyne	

7. Complete the following table by providing the IUPAC names of the following compounds.

	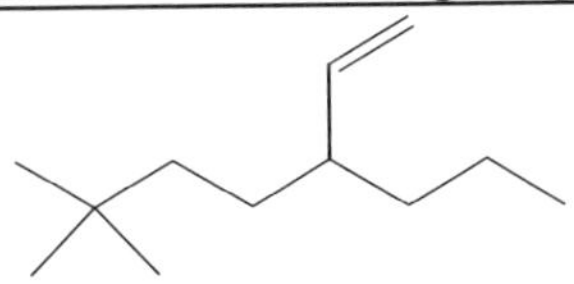 (structure shown)

55

	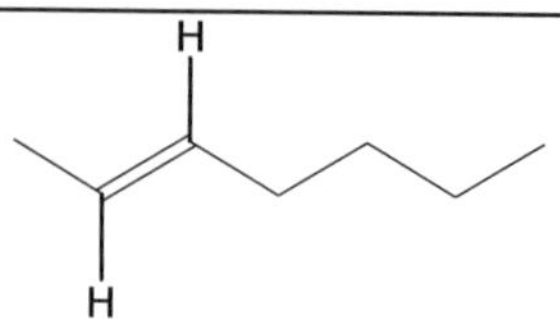 (structure)
	 (structure)
	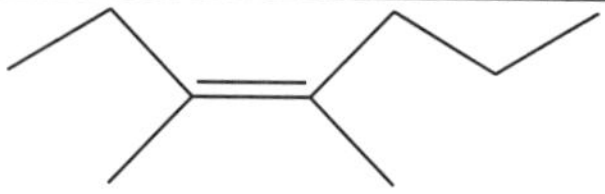 (structure)
	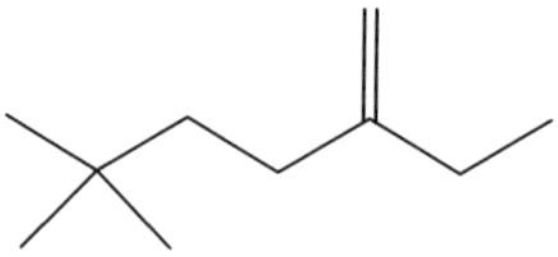 (structure)
	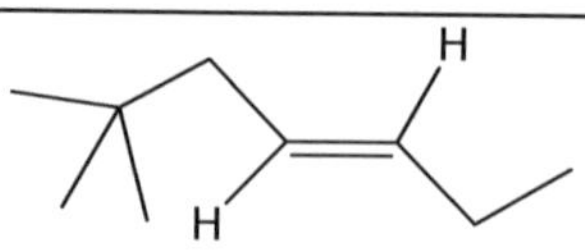 (structure)

8. Which of the following is the best description of a carbon-carbon double bond?

   a) A double bond is made by side to side overlap of atomic *p* orbitals.
   b) A double bond is composed of a σ and a π bond.
   c) A double bond consists of two bonds of equal strength.
   d) Carbon atoms of the double bonds have bent geometry.

9. An unsaturated compound with two rings and no triple bonds has a molecular formula C_9H_{14}. How many double bonds does the molecule have?

   a) 0
   b) 1
   c) 2
   d) 3

10. A hydrocarbon with twelve carbon atoms has two rings and a double bond. How many hydrogen atoms does the molecule have?

    a) 12
    b) 18
    c) 20
    d) 24

Exercises in Organic Chemistry

Answers to Additional Exercises: Questions 6-10

6. Complete the following table by providing the structural formulas of the following compounds.

3-methylcyclopentene	
(Z)-4-bromo-2-pentene	
3,3-dimethylcyclopentene	
4-bromo-2-pentyne	
1-bromo-1-pentyne	

7. Complete the following table by providing the IUPAC names of the following compounds.

6,6-dimethyl-3-propyl-1-heptene	
(E)-2-heptene	
(Z)-2-hexene	
(Z)-3,4-dimethyl-3-heptene	
3-ethyl-5,5-dimethyl-1-hexene	
(E)-6,6-dimethyl-3-heptene	

8. b); 9. b); 10. c).

Veljko Dragojlovic

Exercises in Organic Chemistry

Test 2: CHAPTERS 5-8

1. Which of the following is the strongest acid?

   a) $CH_3CH_2CO_2H$
   b) $ClCH_2CH_2CO_2H$
   c) $ClCH_2CO_2H$
   d) CH_3CO_2H

2. Which of the following carboxylic acids has the largest K_a value?

   a) CH_3CO_2H
   b) ICH_2CO_2H
   c) CF_3CO_2H
   d) $CH_3(CH_2)_2CO_2H$

3. An inductive effect is:

   a) the displacement of electrons along a π bond
   b) the displacement of electrons along a σ bond
   c) the transfer of electrons from a π bond to a σ bond
   d) the transfer of electrons across a benzene ring

4. Stability of the following carbocations is:

   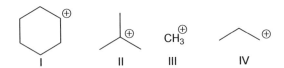

   a) III>IV>I>II
   b) III>IV>II>I
   c) I>II>IV>III
   d) II>I>IV>III

5. Stability of the following free radicals is:

   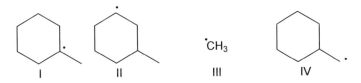

   a) I>II>IV>III
   b) III>IV>II>I
   c) I>II>III>IV
   d) III>IV>I>II

6. Which compound has the lowest heat of hydrogenation?

   a) cis-2,3-dimethyl-3-hexene
   b) trans-2,3-dimethyl-3-hexene
   c) cis-3,4-dimethyl-3-hexene
   d) trans-3,4-dimethyl-3-hexene

7. Crotonic acid, $CH_3CH=CHCO_2H$, a compound found in Texas clay and formed by the dry distillation of wood, exists as the (E)- isomer. Draw a correct geometric representation of this molecule.

8. Write all the isomeric compounds with the formula C_5H_{10}. (Hint: Calculate the Degree of Unsaturation first!). Make sure that you do not include the same isomer twice.

9. Draw the structural formula (including all the hydrogens) for each of the following compounds:

   a) 2-methyl-1-pentene
   b) 3-iodo-1-butyne
   c) 3-methylcyclohexene

10. A compound with a molecular formula $C_{10}H_{16}$ does not react with hydrogen in the presence of platinum catalyst. How many rings does the molecule have?

11. Provide a mechanism for addition of bromine to an alkene.

12. Complete the following table giving the structure or name for each molecule.

Name	Structure
3-methyl-1-pentyne	
acetylene	

13. Draw the structure of the missing compound or reagent in each case:

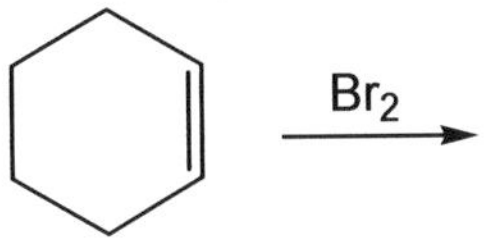

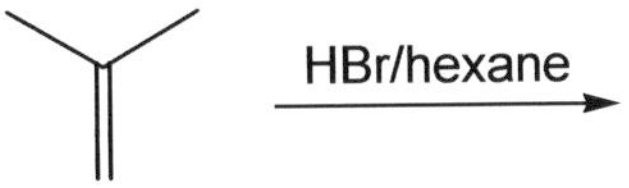

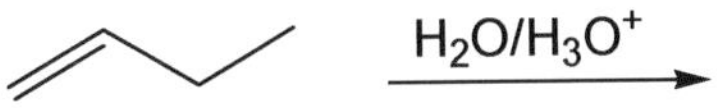

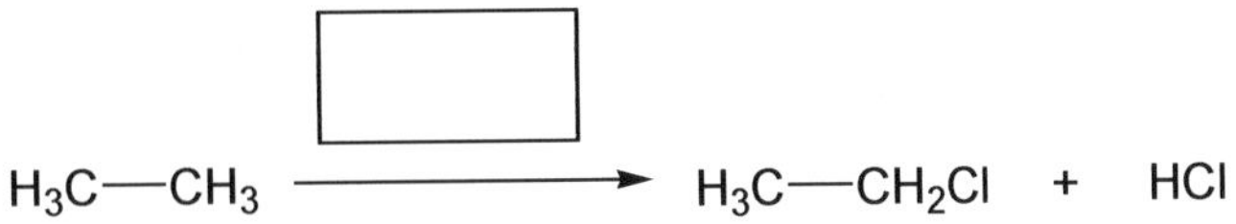

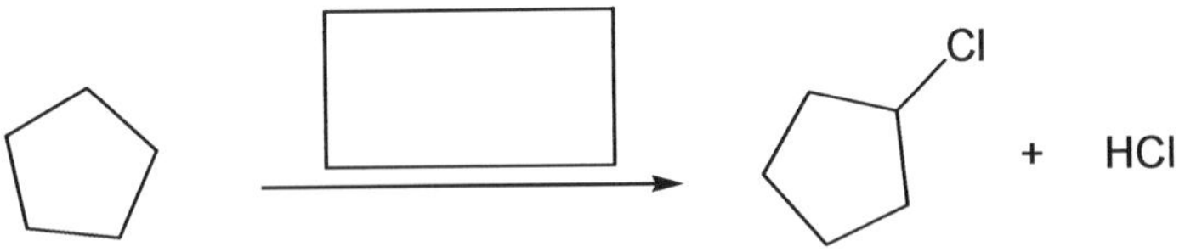

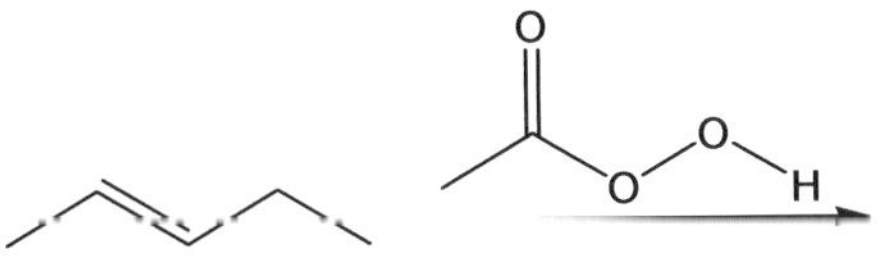

Exercises in Organic Chemistry

[Cyclohexene] →(Br₂)

[2-methylpropene] →(HBr/hexane)

[1-butene] →(H₂O/H₃O⁺)

H₃C—CH₃ →[ ]→ H₃C—CH₂Cl + HCl

[vinylcyclohexane] →[ ]→ [cyclohexyl-CH₂CH₂Br]

[cyclopentane] →[ ]→ [chlorocyclopentane] + HCl

[pentene] →(CH₃C(O)OOH)

14. Draw an energy diagram for a two step reaction passing through an intermediate that is less stable than both the starting material and the product. The product is more stable than the starting material and the first step is the rate determining step.

15. If a reaction has ΔH= -6 kcal/mol, is it exothermic or endothermic? Is it likely to be fast or slow at room temperature? Explain.

Practice Test 2 Solutions

1. c); 2. c); 3. b); 4. a); 5. b); 6. d);
7. (*E*)-Crotonic acid:

8. The compound has: DoU = [(2 × 5 + 2) − 10] / 2 = 1. Therefore, it has either one ring or one double bond.

9. Structural formulas:
a)   b)   c)

10. The compound has three degrees of unsaturation (DoU = [(2 × 10 + 2) − 16] / 2 = 3). Since there is no reaction with hydrogen there are no double or triple bonds. Therefore, the molecule has 3 rings.

11. A mechanism for addition of bromine to an alkene.

Exercises in Organic Chemistry

12. Complete the following table giving the structure or name for each molecule.

Name	Structure
allyl *t*-butyl ether	
5-methyl-2-hexyne	
3-methyl-1-pentyne	
(*E*)-4-methyl-2-hexene	
acetylene	H—C≡C—H
(*Z*)-3-methyl-2-pentene	

13. Draw structure of the missing compound in each case:

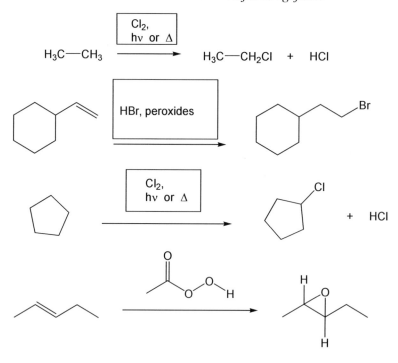

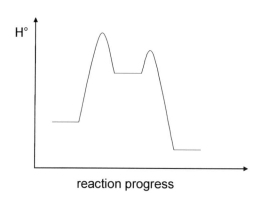

14. Energy diagram.

15. The reaction is exothermic as indicated by the negative sign for enthalpy. Magnitude and sign of enthalpy do not provide any information about the reaction kinetics. Therefore, from the information provided one cannot make a conclusion about the reaction rate at room temperature. To do that one needs information about the activation energy.

Exercises in Organic Chemistry

9. Reactions of Alkenes

Topics to review: mechanism of electrophilic addition to alkenes, free radical addition, regioselectivity, stability of carbocations and free radicals, catalytic hydrogenation reactions and stability of alkenes.

Questions 1-9.

1. An alkene is most likely to react with:

   a) an electrophile.
   b) a nucleophile.
   c) a base.
   d) a Lewis base.

2. Arrange the following carbocations in the order of increasing stability.

3. Arrange the following free radicals in the order of *increasing* stability.

4. Which of the following carbocations is the most stable?

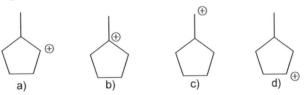

5. Which alkene is the most stable?

   a) *cis*-2,3-dimethyl-3-hexene
   b) *trans*-2,3-dimethyl-3-hexene
   c) *trans*-3,4-dimethyl-3-hexene
   d) *cis*-3,4-dimethyl-3-hexene

65

6. Which of the following alkenes is *the least* stable?

   a) 3-methyl-1-pentene
   b) (*E*)-3-methyl-2-pentene
   c) (*Z*)-3-methyl-2-pentene
   d) Can't tell from a molecular formula, one has to determine heat of hydrogenation first.

7. Which alkene has the lowest heat of hydrogenation?

   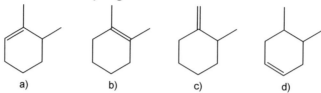

   a)   b)   c)   d)

8. Which alkene has the lowest heat of combustion?

   a) *cis*-2,3-diiodo-2-pentene
   b) *trans*-2,3-diiodo-2-pentene
   c) *cis*-3,4-diiodo-2-pentene
   d) *trans*-3,4-diiodo-2-pentene

9. A substituted carbocation is stabilized by electron delocalization through hyperconjugation. Electron withdrawing groups are not electron donors and, in fact, may destabilize a neighboring carbocation. Predict regioselectivity of the following reaction:

Answers to questions 1-9.

1. This is a closed question. An alkene is a Lewis base, or a nucleophile. Therefore, it reacts with Lewis acids, or electrophiles and the correct answer is a). An alkene also reacts with free radicals, oxidizing agents, dienes, etc.

2. This is a straightforward question. As long as you know that the stability of carbocations increases with an increase in degree of substitution and are able to recognize primary secondary and tertiary carbocation, this is an easy question. Carbocation I is primary, II is secondary, III is methyl and IV is tertiary. Therefore, the order of stability is: III<I<II<IV.

3. As with carbocations, stability of free radicals increases with increase in degree of substitution. Again one needs to correctly identify primary, secondary and tertiary species. Free radical I is primary, II methyl, III tertiary and IV secondary and the order of stability is: II<I<IV<III.

4. This question is similar to question 2. Some students are confused by cyclic compounds. One should treat them the same way as open chain compound. Since the question is which one is the most stable, one should choose the most substituted carbocation and it is the carbocation b).

5. This is a closed question and is relatively straightforward. There are two factors to consider: degree of substitution (the more substituted alkene is the more stable one) and geometry (*trans* isomer is more stable than *cis*). It is the best to draw the individual compounds and compare them. Both 2,3-dimethyl-3-hexenes have trisubstitued double bond, while the two 3,4-dimethyl-3-hexenes have tetrasubstituted double bonds. Therefore, the latter two are more stable. Between the two the *trans* isomer is more stable than the *cis* isomer and the correct answer is c) *trans*-3,4-dimethyl-3-hexene.

6. This is another closed question. The only trick here is to realize that the question asks which compound is the least stable, not the most stable. Again it is the best to draw structural formulas of the compounds and compare them. The reasoning is now reverse from the previous question when we were looking for the most stable alkene. The least substituted one, or a) 3-methyl-1-pentene, is the correct answer. Answer d) is a typical distractor. It is correct that heat of hydrogenation would provide information about the stability of alkenes and if one had the values for the four compounds, one could easily determine which one is the least stable. It is also correct that one can't tell from molecular formula. However, you are provided with names that allow you to draw structural formulas and from them you can tell which one is the most stable.

7. This is a closed question. First, one has to identify what the actual question is. From the provided information one cannot estimate the actual values for heats of hydrogenation of individual compounds. However, to answer this question one does not need to know the absolute values. Rather, one needs to know the relative order of heats of hydrogenation of the compounds above. Therefore, we need to consider what heat of hydrogenation of a molecule tells us. It provides information about the stability of a double bond – the lower the heat of hydrogenation, the more stable the double bond is. Thus, the most stable compound has the lowest heat of hydrogenation and the least stable the highest. Therefore, question about the lowest heat of hydrogenation is in reality a question about the most stable double bond. Stability of a double bond depends on the degree of substitution – the more substituted double bond is the more stable one. Among the four compounds, one with the most substituted double bond is b) 1,2-dimethyl-1-cyclohexene and that is the correct answer.

8. This question is similar to the previous one. The lowest heat of combustion means the most stable double bond and, as already mentioned, the most stable double bond is the most substituted one. Among the four compounds, both compounds c) and d) have tetrasubstituted double bonds. When the degree substituents is the same, the more stable double bond is the one in which the larger substituents are *trans* to each other. That is the compound b) *trans*-2,3-diiodo-2-pentene.

9. The major product is a result of an anti-Markovnikov addition. Two trifluoromethyl group would destabilize the neighboring carbocation making it less stable than the primary carbocation. Therefore, hydrogen adds to the carbon 2.

Additional Exercises: Questions 10-24

10. Predict the major product of each of the following reactions:

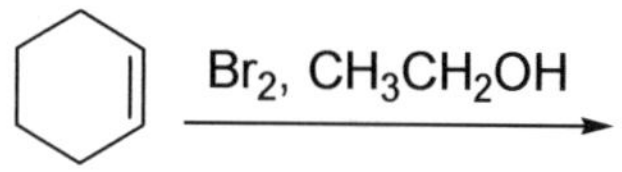

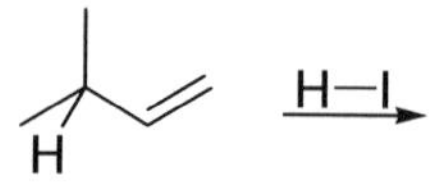

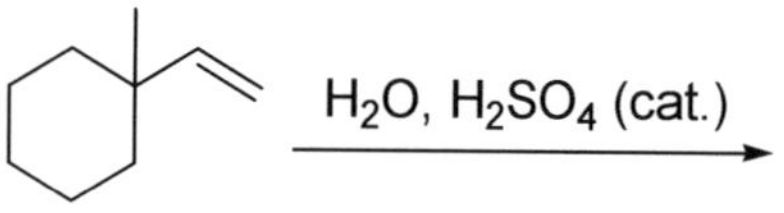

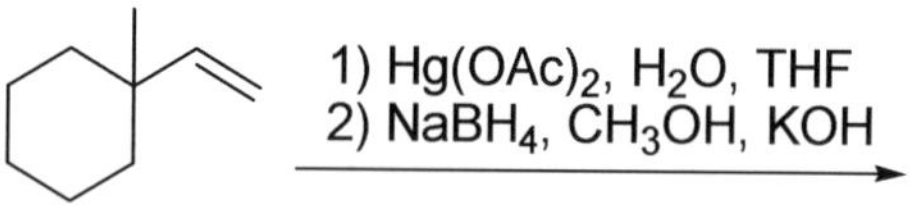

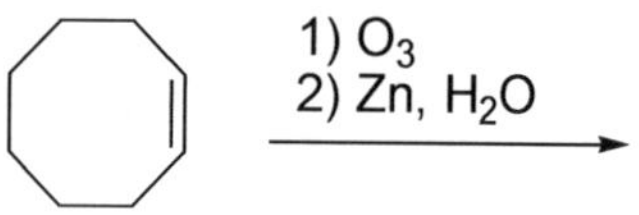

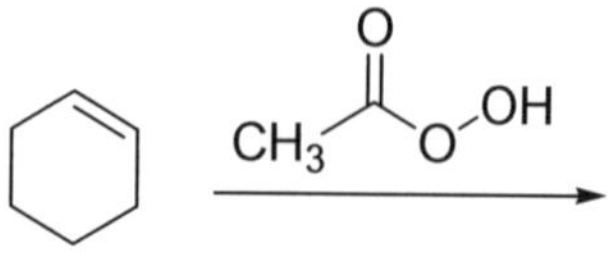

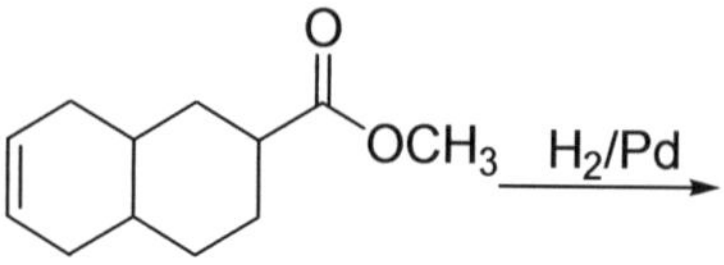

cyclohexene + Br$_2$, CH$_3$CH$_2$OH →

3-methyl-1-butene (CH$_2$=CH–CH(CH$_3$)$_2$) + H–I →

1-vinyl-1-methylcyclohexane + H$_2$O, H$_2$SO$_4$ (cat.) →

1-vinyl-1-methylcyclohexane
1) Hg(OAc)$_2$, H$_2$O, THF
2) NaBH$_4$, CH$_3$OH, KOH →

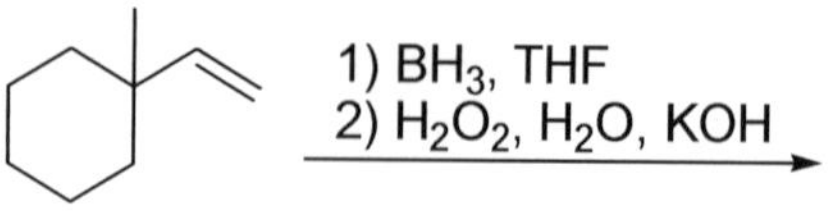

1-vinyl-1-methylcyclohexane
1) BH$_3$, THF
2) H$_2$O$_2$, H$_2$O, KOH →

cyclooctene
1) O$_3$
2) Zn, H$_2$O →

cyclohexene + CH$_3$C(O)OOH →

(octahydronaphthalene with alkene and –C(O)OCH$_3$ substituent) + H$_2$/Pd →

Exercises in Organic Chemistry

11. What was the starting material in of each of the following reactions:

? $\xrightarrow{Cl_2, CH_2Cl_2}$ Ph-CHCl-CHCl-Ph

? $\xrightarrow{Br_2, CH_2Cl_2}$ 1,2-dibromocyclooctane

? $\xrightarrow{HCl, CH_2Cl_2}$ CH$_3$-CHCl-CH$_2$-CH$_2$-C(=O)-OCH$_3$

? $\xrightarrow{CH_3OH, H_2SO_4 \text{ (cat.)}}$ cyclopentyl-OCH$_3$

? $\xrightarrow[\text{2) H}_2\text{O}_2]{\text{1) O}_3}$ HO-C(=O)-(CH$_2$)$_5$-C(=O)-OH

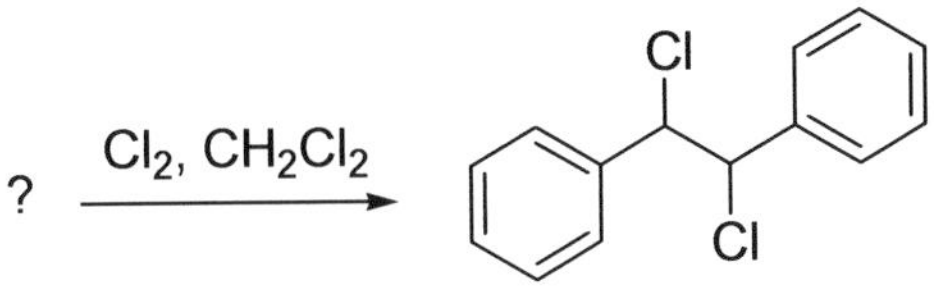

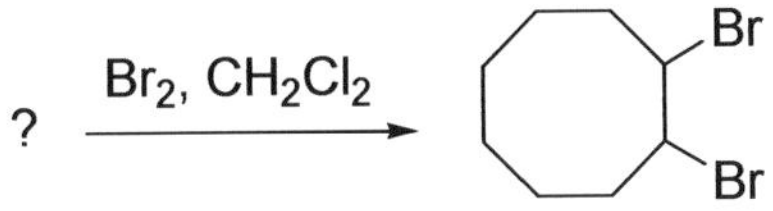

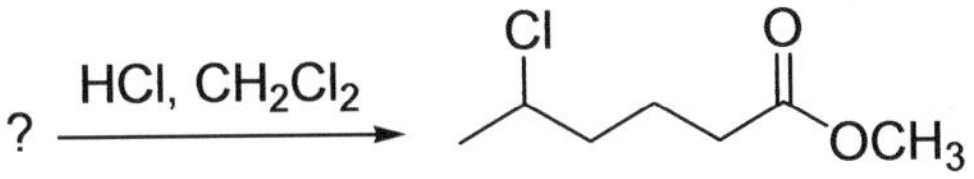

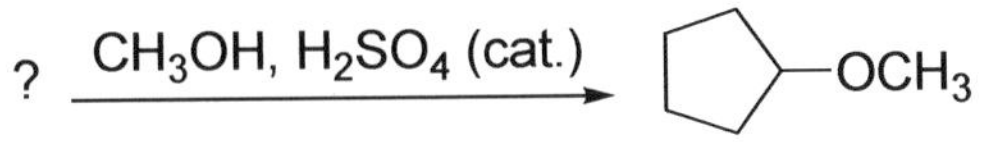

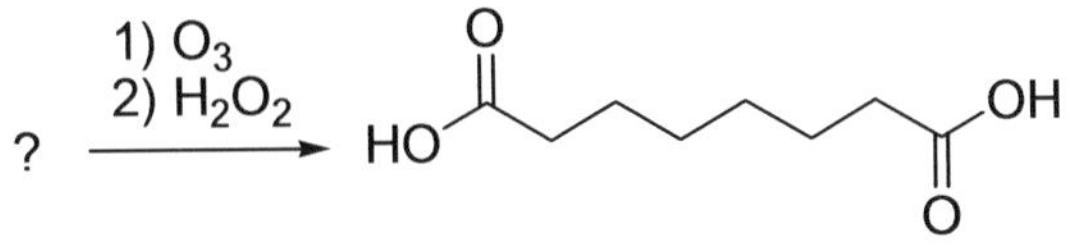

12. Provide the reagent for each of the following reactions:

cyclohexene $\xrightarrow{?}$ HO-C(=O)-(CH$_2$)$_4$-C(=O)-OH

vinylcyclopentane $\xrightarrow{?}$ cyclopentyl-CH$_2$-CH$_2$-OH

methylenecyclohexane $\xrightarrow{?}$ 1-iodo-1-methylcyclohexane

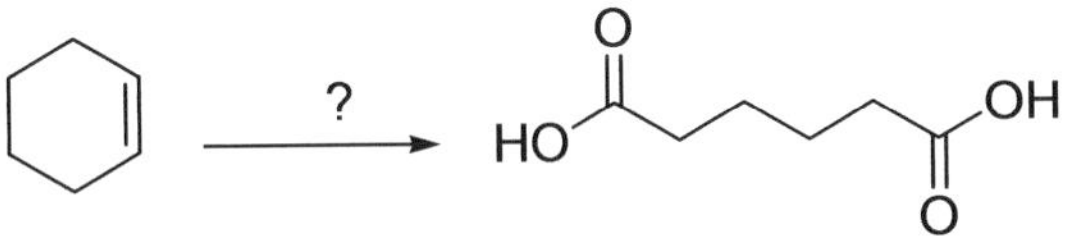

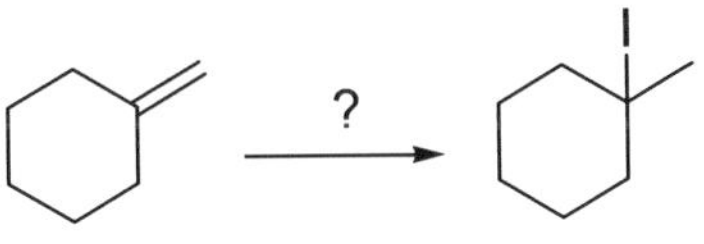

13. Which of the following additions of HCl to an alkene does not obey the Markovnikov rule?

(a) $(CH_3)_2C=CH_2$
(b) $C_6H_5CH=C(CH_3)_2$
(c) $CH_3CH=CH_2$
(d) $CH_3CH_2CH=CHCH_3$

14. Which carbocation is the major product of the reaction shown below?

$$(CH_3)_2C=CHCH_3 + H^+ \rightarrow$$

   a) $(CH_3)_2C^+-CH_2CH_3$
   b) $(CH_3)_2CH-C^+CH_3$
   c) $(CH_3)_2C-CH_2CH_2^+$
   d) $CH_3CH^+-C(CH_3)CH_3$

15. Ethene is more reactive towards bromine than ethane because:

   a) Ethene has a shorter carbon-carbon bond than ethane.
   b) Ethene is a nucleophile while ethane is not.
   c) Ethene is less sterically hindered than ethane.
   d) The heat of combustion of ethene is greater than that of ethane.

16. What is the major product in a reaction of 2-butene with bromine in the presence of an excess of sodium iodide?

   a) 2,3-dibromobutane
   b) 2,3-diiodobutane
   c) 2-bromo-3-iodobutane
   d) 1-bromo-4-iodo-2-butene

17. What is the major product in a reaction of ethene with bromine in methanol as a solvent?

   a) 1,2-dibromoethane
   b) 1,2-dimethoxyethane
   c) 1,1,2-tribromoethane
   d) 1-bromo-2-methoxyethane

18. Propene reacts with hydrogen chloride faster than ethene because:

   a) propene is less sterically hindered.
   b) propene has a shorter double bond than ethene.
   c) propene gives the more stable intermediate free radical than ethene.
   d) propene gives the more stable intermediate carbocation than ethene.

19. Which of the following reactions is carried out in the presence of a heterogeneous catalyst?

   a) hydration of an alkene
   b) hydrogenation of an alkene
   c) mercuric ion catalyzed hydration of an alkene
   d) addition of HBr in the presence of peroxides

20. Oxymercuration-demercuration of an alkene follows Markovnikov rule because:

    a) addition of water results in breaking of the weaker carbon-mercury bond.
    b) the most stable carbocation is the reaction intermediate.
    c) the reaction proceeds via a mercurinium intermediate.
    d) addition is governed by steric factors.

21. Which of the following statements is not correct?

    a) Tertiary free radicals are more stable than secondary.
    b) Tertiary carbocations are more stable than secondary.
    c) Primary free radicals are less stable than secondary.
    d) Secondary carbocations are less stable than primary.

22. Which alkene has the highest heat of hydrogenation?

    a) cis-2,3-dimethyl-3-hexene
    b) trans-2,3-dimethyl-3-hexene
    c) cis-3,4-dimethyl-3-hexene
    d) trans-3,4-dimethyl-3-hexene

23. Alcohols are reaction products of ozonization of an alkene when the workup is performed with:

    a) zinc and water
    b) hydrogen peroxide
    c) dimethyl sufide
    d) sodium borohydride

24. Carbon dioxide is one of the reaction products of permanganate oxidation of:

    a) 1-pentene
    b) cis-2-pentene
    c) cyclopentene
    d) 1-methylcyclopentene

Answers to Additional Exercises: Questions 10-24

10. Reaction products:

11. Starting materials:

PhCH=CHPh →(Cl₂, CH₂Cl₂)→ PhCH(Cl)CH(Cl)Ph

cyclooctene →(Br₂, CH₂Cl₂)→ 1,2-dibromocyclooctane

methyl pent-4-enoate →(HCl, CH₂Cl₂)→ methyl 4-chloropentanoate

cyclopentene →(CH₃OH, H₂SO₄ (cat.))→ methoxycyclopentane

cyclooctene →(1) O₃; 2) H₂O₂)→ HOOC-(CH₂)₆-COOH (suberic/octanedioic acid)

12. Reagents:

cyclohexene →(1) O₃, CH₂Cl₂; 2) H₂O₂, H₂O)→ HOOC(CH₂)₄COOH

Other reagents such as KMnO₄/H₃O⁺ also work.

cyclopentyl-CH=CH₂ →(1) BH₃, THF; 2) H₂O₂, H₂O, NaOH)→ cyclopentyl-CH₂CH₂-OH

methylenecyclohexane →(HI)→ 1-iodo-1-methylcyclohexane

13. b); 14. a); 15. b); 16. c); 17. d); 18. d); 19. b); 20. a); 21. d); 22. c); 23. d); 24. a).

Veljko Dragojlovic

Exercises in Organic Chemistry

10. Reactions of Alkynes

Topics to review: mechanism of addition to alkynes, free radical addition, regioselectivity, stability of carbocations and free radicals, catalytic hydrogenation reactions, dissolving metal reductions and reactions of terminal alkynes.

1. What is the starting material A in the following reaction?

    $$A + 2H_2 \xrightarrow{Pt} B$$

    a)  cyclohexane   b) 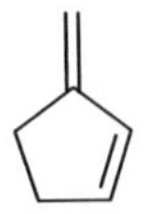 methylenecyclopentane   c) 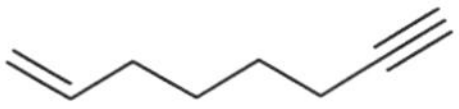 hept-1-en-6-yne   d) 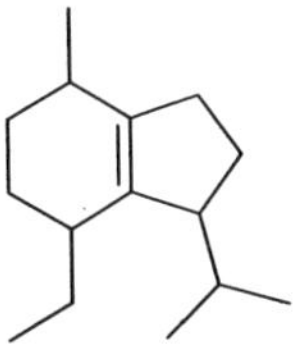 bicyclic structure with ethyl and isopropyl substituents

2. An eight-member ring is the smallest ring that can accommodate either a triple bond or a *trans* double bond. Provide the formulas of the products in the reactions below.

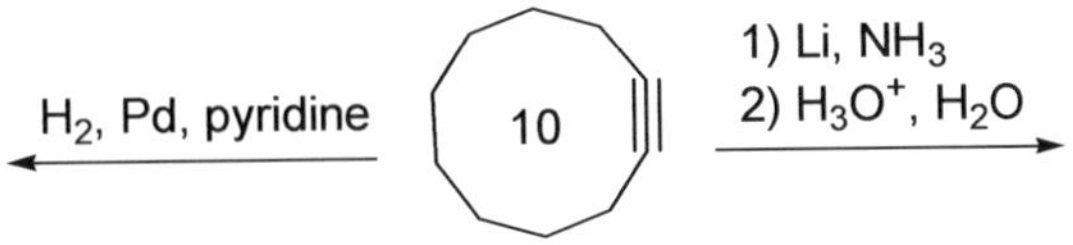

    cyclodecyne (10-membered ring with triple bond):
    - H_2, Pd, pyridine ←
    - → 1) Li, NH_3 2) H_3O^+, H_2O

3. There are two possible starting materials to prepare 2,2-dibromobutane from an alkyne and an excess of HBr and only one way to prepare 2,2-dibromopentane from an alkyne and an excess HBr. Provide the structures of the alkynes.

4. Predict the major product of each of the following reactions:

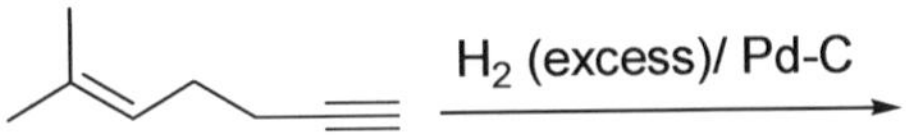

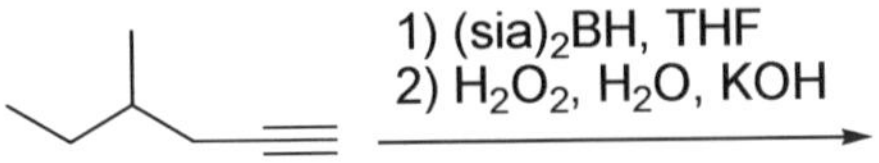

    (4-methylpent-1-en... alkyne) + H_2 (excess)/ Pd-C →

    (4-methylpent-1-yne)
    1) $(sia)_2BH$, THF
    2) H_2O_2, H_2O, KOH

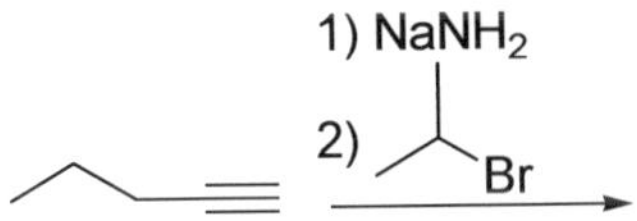

    (terminal alkyne)
    1) $NaNH_2$
    2) 2-bromo-2-methylpropane (tert-butyl bromide)

5. Provide the final products of these multi-step reactions.

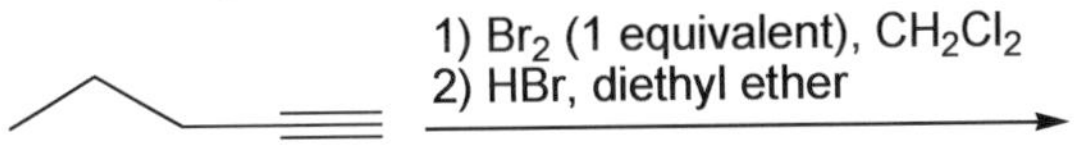

    (pent-1-en-4-yne type):
    1) Br_2 (1 equivalent), CH_2Cl_2
    2) HBr, diethyl ether

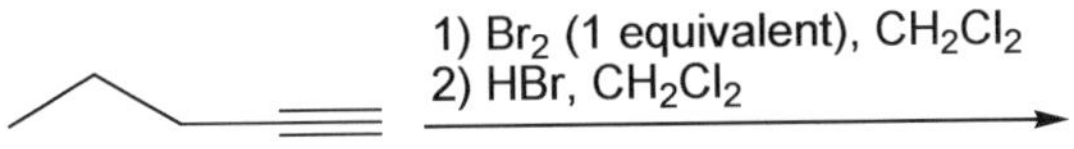

    (pent-1-en-4-yne type):
    1) Br_2 (1 equivalent), CH_2Cl_2
    2) HBr, CH_2Cl_2

75

6. Provide reagents for the following reactions:

   [structure: enyne → diene] ?

   [structure: enyne → ene-yne (internal alkyne)] ?

   [structure: alkyne → ketone] ?

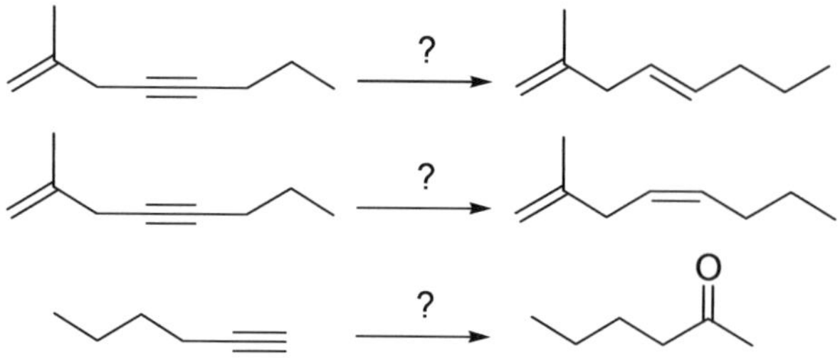

7. Provide starting materials for the following reactions:

   ? $\xrightarrow{\text{HBr (xs), CH}_2\text{Cl}_2}$ [cyclooctane with geminal Br, Br]

   ? $\xrightarrow[\text{2) H}_2\text{O}]{\text{1) O}_3\text{, CH}_2\text{Cl}_2}$ [(CH$_3$)$_2$C(OH)COOH] + HO–C(=O)–CH(CH$_3$)$_2$

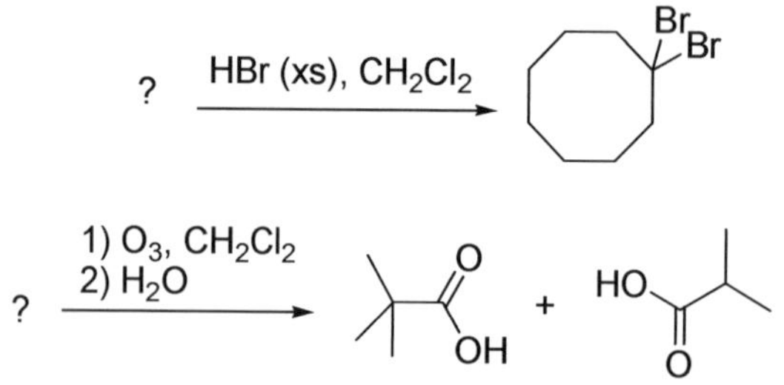

8. 1-Butyne and 2-butyne may be distinguished by reactions with which of the following reagents?

   a) H$_2$/Pd/pyridine
   b) Na/NH$_3$
   c) NaNH$_2$
   d) HBr

9. Reaction of 1-hexyne with disiamylborane followed by an oxidation gives:

   a) a primary alcohol.
   b) a secondary alcohol.
   c) an aldehyde.
   d) a ketone.

10. In a reaction of 2-hexyne with permanganate at a room temperature, the major product is:

    a) 2,3-hexadiol
    b) 2,3-hexadione
    c) acetic and butanoic acids
    d) ethanal and butanal

11. What is the major product in a reaction of 1-butyne with an excess of hydrogen bromide in dichloromethane?

    a) 2,2-dibromobutane
    b) 1,2-dibromobutane
    c) 1,1-dibromobutane
    d) 1,1,2,2-tetrabromobutane

Answers

1. This is a closed question. In a catalytic hydrogenation reaction a molecule of hydrogen reacts with a π bond. In this particular case, the molecule A reacted with 2 equivalent of hydrogen, which means that it has 2 π bonds. In turn, it means that compound A has either 2 double bonds or a triple bond. Now we should consider the four possible answers.

a) Cyclohexene is the only of the four compounds that has 2 degrees of unsaturation. Hydrogen in the presence of a noble metal catalyst reacts with π bonds. In this case, only one of the two unsaturations is a π bond, the other one is a ring. Therefore, this is a distractor designed for students who recognize that it has two degrees of unsaturation, but do not know that only π bonds undergo hydrogenation.

b) This compound contains two π bonds in form of two carbon-carbon double bonds. This is the correct answer.

c) A carbon-carbon triple bond reacts with 2 equivalent of hydrogen in the presence of a noble metal catalyst. However, this compound also has a carbon-carbon double bond. Therefore, it would be expected to react with a total of three equivalents of hydrogen. It is another distractor. Students who know that an alkyne reacts with 2, but do not realize that at the same time the alkene functional group reacts with an additional equivalent of hydrogen, may go for an answer like this.

d) This is a compound of a more complex structure. Since it has only one double bond it is not a correct answer. It is another distractor. Students with little, or no knowledge, may choose the most complex answer, like this one.

2. Both reagents reduce an alkyne to an alkene. The difference is that catalytic hydrogenation results in a *syn* addition (addition from the same side) while dissolving metal reduction results in a *trans* addition of hydrogen. Therefore the products are:

Although the eight-member ring is the smallest ring that accommodate a *trans*-double bond, it is only in rings with ten or more carbon atoms that the *trans* isomer is the major product. Due to a high angle strain, eight- and nine-member rings give predominantly *cis* alkene. Remember that dissolving metal reduction gives the most stable alkene.

3. Starting materials:

4. Reaction products:

(reactions shown with structures)

- Alkene-alkyne + H₂ (excess)/ Pd-C → saturated branched alkane
- 3-methyl-pent-1-yne type + 1) (sia)₂BH, THF; 2) H₂O₂, H₂O, KOH → aldehyde
- Terminal alkyne + 1) NaNH₂; 2) isopropyl bromide → internal alkyne with isopropyl group

5. Products:

- Pent-1-yne + 1) Br₂ (1 equivalent), CH₂Cl₂; 2) HBr, diethyl ether → tribromide (vicinal dibromide with additional Br)
- Pent-1-yne + 1) Br₂ (1 equivalent), CH₂Cl₂; 2) HBr, CH₂Cl₂ → tribromide (geminal/vicinal pattern)

6. Reagents:

- Enyne + Na, NH₃, -78°C → trans-alkene (with isopropenyl group retained)
- Enyne + H₂/Pd/pyridine → cis-alkene
- Terminal alkyne + HgSO₄, H₂O, H₂SO₄ → methyl ketone

7. Starting materials:

- Cyclooctyne + HBr (xs), CH₂Cl₂ → gem-dibromocyclooctane
- 2,2,5,5-tetramethyl-3-hexyne + 1) O₃, CH₂Cl₂; 2) H₂O → pivalic acid + isobutyric acid (HO-C(=O)-CH(CH₃)₂)

8. c; 9. c; 10. b; 11. a.

Exercises in Organic Chemistry

11. Stereochemistry

Topics to review: isomerism, chirality centers, nomenclature of stereoisomers, priority rules, representation of enantiomers, Fischer and Haworth projections, optical rotation, optical purity, racemic mixtures, compounds with more than one chirality center, meso compounds, chirality of cyclic compounds.

Questions 1-5.

1. Classify the following isomers.

2. Draw the other conformational isomer of the amine below.

3. Draw Fischer projections of these two compounds:

4. Draw Haworth projections of the following compounds:

5. Which of the following molecules are chiral and how many chirality centers are present in each of the chiral molecules?

79

Answers to questions 1-5.

1. Classification of isomers:

skeletal (different carbon skeletons)   **functional** (different functional groups)

positional (different positions of functional groups)

2. A molecule of amine has tetrahedral geometry with one corner of tetrahedron being occupied by the nitrogen's free electron pair. Free electron pair can easily flip from "up" to "down" position and therefore, unlike tetrahedral carbon compounds, amines have different conformations – not different configurations. The other conformation of methylethylamine is:

3. There are several ways to correctly represent the compounds by means of Fischer projections. One of the Fischer projections for each compound is shown below:

4. Haworth formulas represent each ring as being planar. The ring is drawn in a horizontal plane and bonds on each ring atom as being vertical. Although we know that this is not an accurate representation of a cyclic compound, Haworth's formulas are helpful in study of stereochemistry of ring compounds – it is easy to spot *cis* and *trans* isomers, chiral compounds, symmetry elements and the like. There are several ways to correctly represent the compounds by means of Haworth formulas. Some of them are shown on the left.

Exercises in Organic Chemistry

5. Chirality and chirality centers (indicated with *):

chiral   not chiral   chiral

chiral   chiral

Questions 6-10.

6. Redraw each of the following structural formulas in such a way that the lowest priority substituent points away from the viewer.

7. How many stereoisomers of lanosterol are possible?

8. How many stereoisomers are possible in the following molecule?

$$PhCH(Cl)CH(Ph)CH(Cl)Ph$$

   a) three
   b) four
   c) five
   d) eight

9. How many pairs of enantiomers are possible in the following molecule?

$$CH_3CHClCHBrCHClCH_2Cl$$

   a) two
   b) four
   c) six
   d) eight

10. Provide formulas of all nine isomers of 1,2,3,4,5,6-hexamethylcyclohexane.

Answers to questions 6-10.

6. Vertical bonds point away from the viewer. Therefore, any correct representation in which the lowest priority substituent points away from the viewer is acceptable. One should first correctly assign priority to the substituents and next apply allowed transformations to the Fischer formulas to end up with a formula in which the lowest priority substituent points away from the viewer. There are a number of correct solutions and one for each formula is shown below.

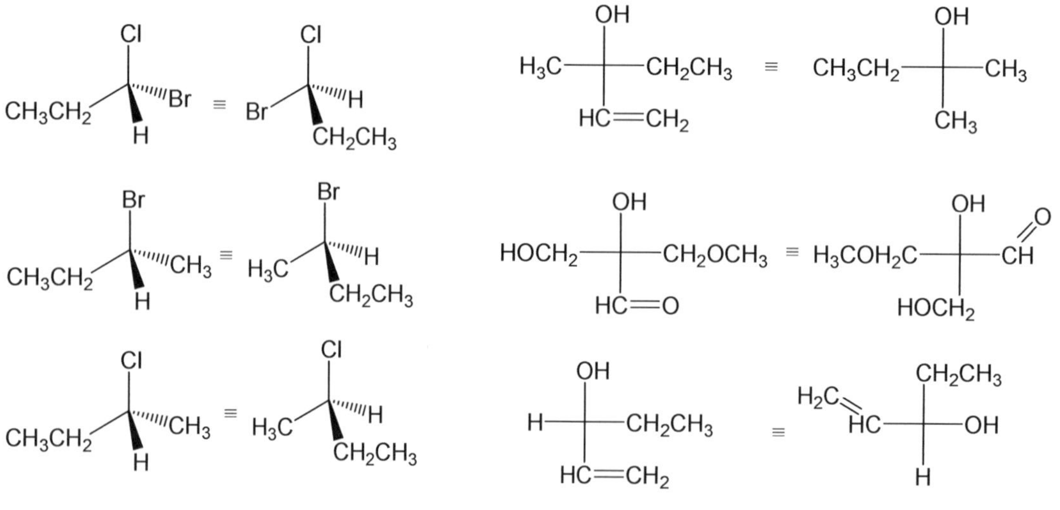

7. A maximum number of stereoisomers is given by the formula 2^n, where n is the number of chirality centers in the molecule. Therefore, the first step is to determine the number of chirality centers in the molecule of lanosterol. You should practice to do that on a line formula of a compound. However, if you are not able to do that, you should draw either a condensed or a complete structural formula and determine the number of chirality centers. With a large molecule, such as lanosterol, utility of line formula becomes apparent. During a test, it may take considerable time to draw a complete structural

formula of a large molecule – that is why you should practice identifying chirality centers from a line formula. Chirality centers are shown below.

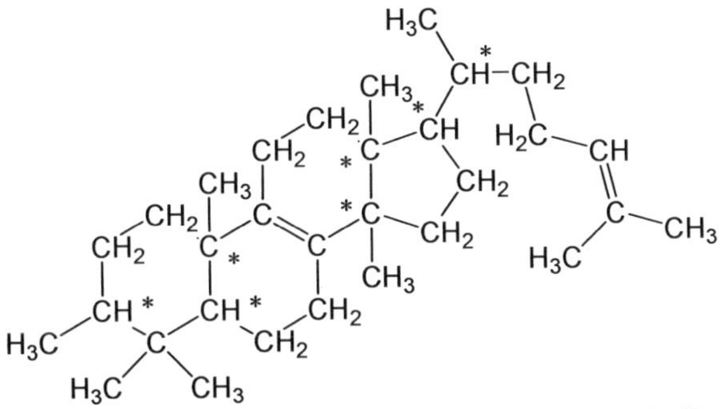

Here is a condensed structural formula with chirality centers indicated.

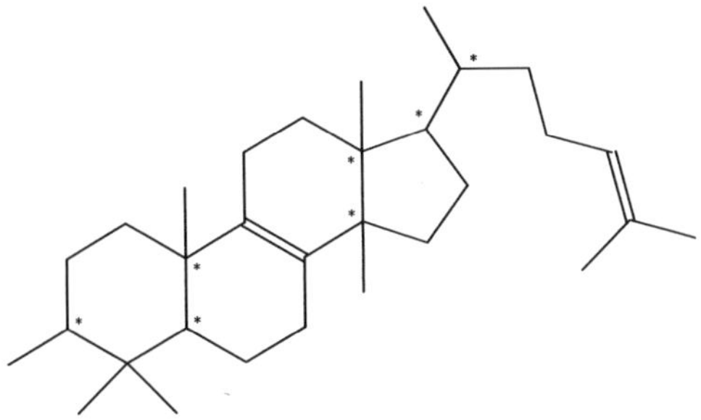

Therefore, lanosterol has a total of seven chirality centers and the maximum number of stereoisomers is $2^7 = 128$. Since this is the maximum number of stereoisomers, the final step is to check whether the molecule has a plane of symmetry, which would reduce the number of stereoisomers. Molecule of lanosterol has no plane of symmetry and, therefore, the number of stereoisomers is 128.

8. By now you know that you need to determine the number of chirality centers first. Here is the formula with labeled stereocenters.

$$PhC^*H(Cl)C^*H(Ph)C^*H(Cl)Ph$$

If you have difficulty identifying them, you may want to use a Fischer projection. On an acyclic molecule, Fischer projections are the most convenient formulas to identify chirality centers.

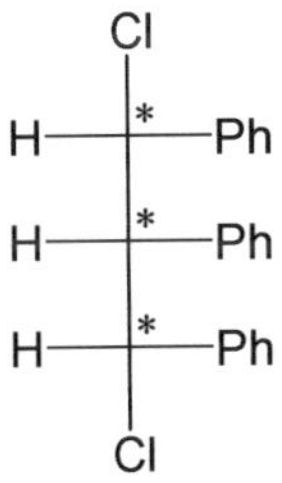

Since there are three chirality centers, the maximum number of stereoisomers is eight. However, the molecule has a plane of symmetry and the actual number of stereoisomers is less than eight. You may want to draw all eight, identify identical ones, and find out how many stereoisomers are actually different.

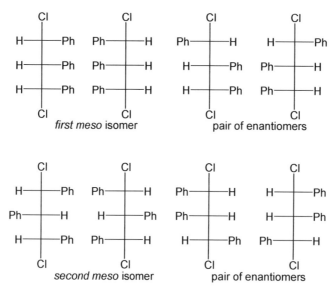

Note that there is only one pair of enantiomers – a 180° rotation, which is an allowed transformation of a Fischer projection formula, converts one pair into another. Therefore, they are identical compounds. Thus, there are two enantiomers and two different *meso* isomers and the total number of stereoisomers is b) four.

9. To answer this question you need to determine the number of chirality centers first. Here is the formula with labeled stereocenters.

$$CH_3C^*HClC^*HBrC^*HClCH_2Cl$$

If you have difficulty identifying them, you may want to use a Fischer projection. On an acyclic molecule, Fischer projections are the most convenient formulas to identify chirality centers.

There are three chirality centers. The molecule does not have a plane of symmetry and the actual number of stereoisomers equals the maximum number of stereoisomers, which is eight ($2^3 = 8$). Here is where a lot of students make a mistake. They go ahead and select answer d) eight. However, the question was: "How many **pairs of enantiomers**... ?" Each pair is composed of two stereoisomers. Therefore, the number of pairs is half the number of stereoisomers and the correct answer is b) four.

10. The easiest way to draw all the different isomers is to use Haworth projections. That way one can easily see stereochemical relationship between the methyl groups. When solving problems like this one has to be careful not to draw the same structure twice. Another important consideration is not to forget that enantiomers are different compounds. In this case there is only one pair of enantiomers (compounds **7** and **8**). Each of the remaining seven compounds has at least one plane of symmetry – try to find them.

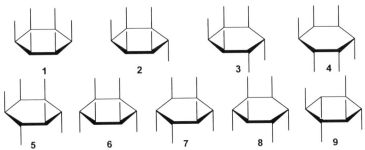

Questions 11-13.

11. Identify the chirality center in limonene, draw the two enantiomers of limonene and assign relative configuration to each enantiomer.

12. The sample of tartaric acid has $[\alpha]_D=+12°$. Explain which of the two formulas is the correct one?

13. Pure (+)-tartaric acid has specific rotation of +12.0°. Calculate optical purity of sample of tartaric acid that has $[\alpha] = -2.4°$.

Answers to questions 11-13.

11. The question consists of three parts:
    (i)   Identify chirality center in limonene.
    (ii) Draw the two enantiomers.
    (iii) Assign relative configuration to each enantiomer.

    A complete answer must address all three parts. First you have to identify the chirality center. If you cannot spot it right away, draw a complete structural formula of limonene (shown below).

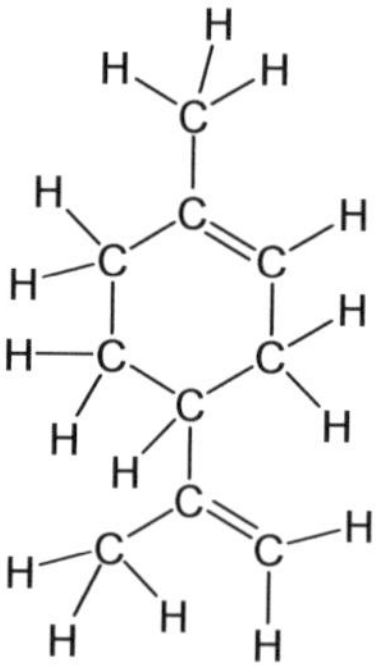

    Can you identify the chirality center now?
    It may be still difficult to identify it. You are looking for a carbon atom with four different substituents. You may have correctly identified the carbon 4 as the chirality center, by considering the other carbon atoms and eliminating them as possible chirality centers either because they had two, or three, identical substituents (hydrogen atoms), or because they were sp^2 hybridized. Do you see why carbon 4 is a chirality center? If you were shown a formula of limonene, would you have realized that it was a chiral molecule? How are the four substituents on the carbon 4 different? 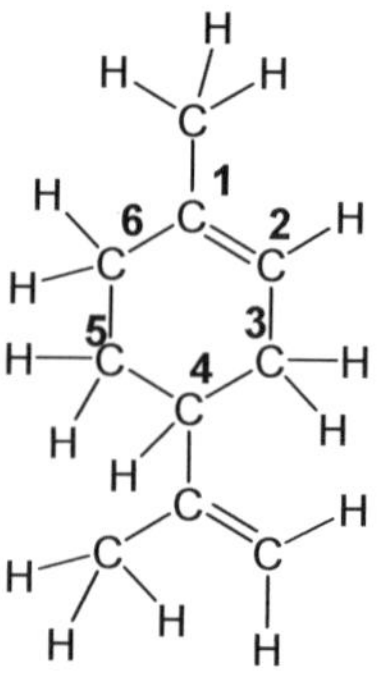
    Two of the exocyclic substituents (those outside the ring) are obviously different. One of them is a hydrogen atom while the other one is a vinylic group. The situation is more difficult with the remaining two substituents, which are parts of the ring. The two ring substituents bonded to carbon 4 are two –CH_2- groups (two methylene groups) and here is where some students get confused. They see two identical substituents, do go any further, and conclude that the atom could not be a chirality center. If a possible chirality center is a part of a ring then the two "halves" of the ring (labeled as "A" and "B") are considered to be the two substituents.
    Here the two ring "substituents" of limonene are labeled as substituent A and substituent B. Substituent A contains a –CH= group (carbon 2) bonded to the methylene, while substituent B contains another methylene group (carbon 6) bonded to the first methylene group. Therefore, the two substituents are clearly different and carbon 4 is a chirality center. As a part of your answer you should indicate the chirality center with a star.

Exercises in Organic Chemistry

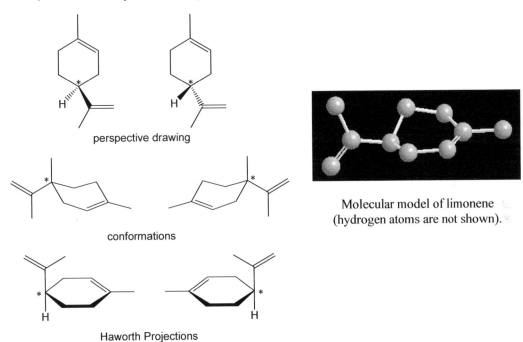

Two ring "substituents" on the carbon 4.

Next, you should draw the two enantiomers. Shown below are several different ways to draw the two enantiomers. To show the two enantiomers, you have to use one of representations that show stereochemistry of the molecule. Some possible representations include perspective drawing, drawing of conformations (note conformation of a cyclohexene ring) and Haworth projection. Although it is possible, we don't usually use Fischer projections to represent a chirality center that is a part of a ring.

perspective drawing

conformations

Molecular model of limonene (hydrogen atoms are not shown).

Haworth Projections

Finally, you have to determine relative stereochemistry of each enantiomer. The best way to do that is from a perspective formula. The lowest priority substituent, hydrogen (labeled as substituent "4"), is easy to identify.

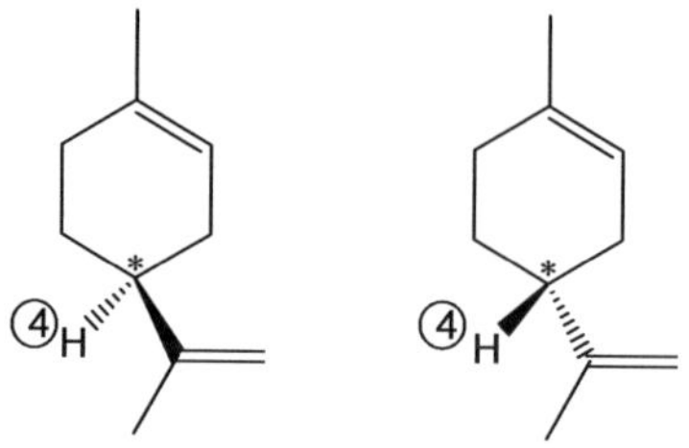

The enantiomer on the left has the lowest priority substituent pointing away from the viewer. Therefore, it is easier to determine its configuration. To determine priorities of the remaining substituents, it is the best to use a complete structural formula. The remaining three substituents are all carbon atoms and, therefore, have the same priority. Since two of the substituents contain double bonds they have to be converted into single bonds by adding virtual atoms. Formula of limonene with added virtual atoms is shown on the right.

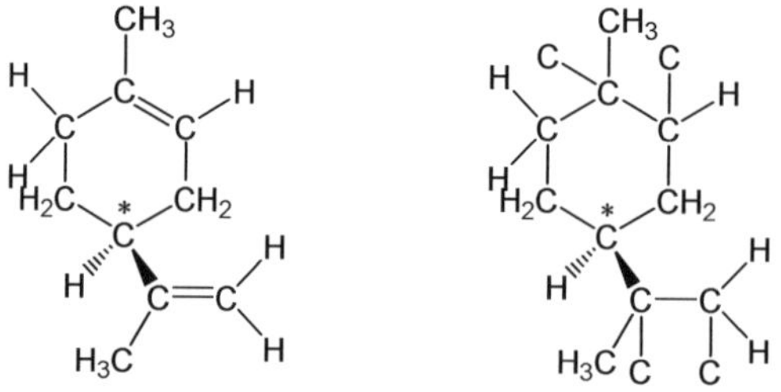

Next, we should determine the highest priority path of each of the three remaining substituents. Note that the ring carbon 1 is common to both ring substituents (shown earlier as substituent A and substituent B) and "doesn't count." There are three carbon atoms bonded to the exocyclic carbon (carbon outside ring), which is attached to the chirality center. They are –CH$_3$, virtual carbon and –CH$_2$ bonded to a virtual carbon. Among the three, the last one is the highest priority path, due to its virtual carbon, and that is the pathway we should follow when comparing that substituent with the other two endocyclic (inside the ring) substituents. The highest priority path for each of the three substituents is shown in bold.

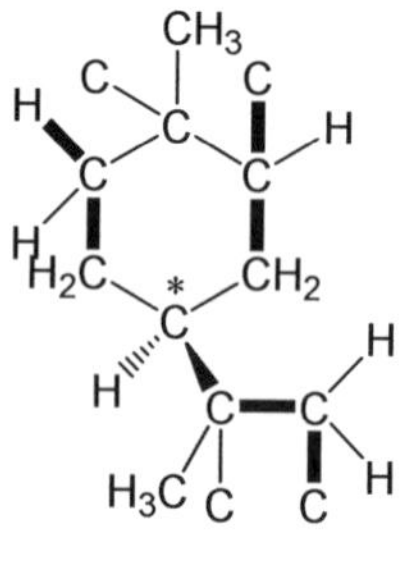

When comparing the three substituents, we first compare the atoms directly bonded to the chirality center. They are all carbon atoms and have equal priority. Each of those three carbons is bonded in turn to another carbon atom and the priorities of the three substituents are still equal. Going further down the highest priority chain of each substituent, we find that the ring substituent to the left (labeled earlier as the substituent B) has a hydrogen atom, while the remaining two substituents have virtual carbon atoms. Since hydrogen has lower priority than carbon (including a virtual carbon), the substituent to the left has lower priority than the other two and has the third highest priority.

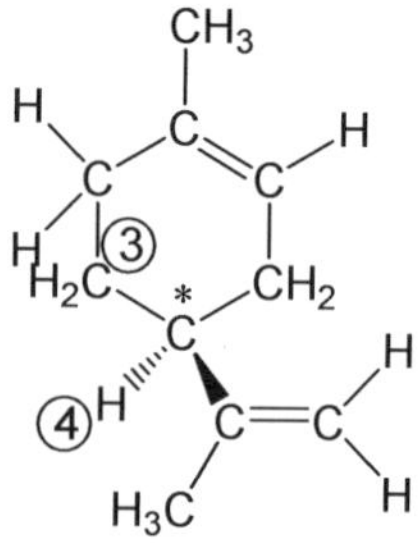

Exercises in Organic Chemistry

We have reached the end of the highest priority pathways of the substituents and there is still a tie between the remaining two substituents. To differentiate between the two we now have to follow the chain of the second highest priority for each of those two substituents. These chains are shown below.

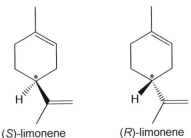

The exocylcic, vinylic, carbon is bonded to another carbon – that of the methyl group, while the endocyclic carbon (substituent A) is bonded to a hydrogen atom. Therefore, vinylic substituent has a higher priority and is the highest priority substituent.

Since the lowest priority substituent already points away from the viewer, we can determine the relative configuration of the molecule. Three order of priorities of substituents makes a counterclockwise array and, therefore, relative configuration of this enantiomer is (S). Conversely, the other enantiomer has (R) configuration.

(S)-limonene   (R)-limonene

12. Those who did not study have no idea that magnitude or sign or rotation cannot be related to the structure of the compound. Therefore, they have no hope of answering a question like this. Even students who are familiar with the topic often find this question confusing as they try to relate the particular value of +12° to the correct answer. Actually, the question is very simple. Only one of the two isomers is chiral! Therefore, the actual question is: "Which of the two compounds is chiral?" Thus, one should identify the chiral compound and it is the threo isomer (I). Erythro isomer (II) has a plane of symmetry and does not rotate a plane of polarized light. It is a meso isomer.

13. Rotation of a solution is due to the presence of an excess of one enantiomer. A mixture of equal amounts of the two enantiomers does not rotate the plane of polarized light. Therefore, optical purity is:

$$(-2.4°/+12°) \times 100\% = -20\%.$$

Note the "-" sign. It means that the other enantiomer, (-)-tartaric acid, is in excess and it is in 20% excess. Therefore, optical purity of a sample of **(-)-tartaric acid** is 20%. In other words, the sample contains 60% (-)-tartaric acid and 40% (+)-tartaric acid.

Additional Exercises: Questions 14-50

14. Draw the other conformation of the amine below.

15. Identify chirality center(s) in the following molecules.

16. How many chirality centers are present in 2-bromonorbornane?

17. Identify with an asterisk (*) all the chirality centers in the following molecule.

18. Draw both **Sawhorse Formulas** and **Perspective Line Formulas** of the two enantiomers of 2-iodobutane.

19. Draw all the stereoisomers of 3,4-dichlorohexane.

Exercises in Organic Chemistry

20. How many pairs of enantiomers are possible in the following molecule?

```
        CHO
    H ──┼── OH
    H ──┼── OH
    HO──┼── H
        CH₂OH
```

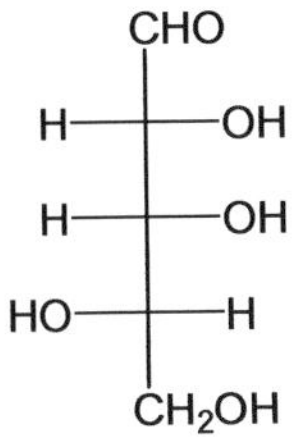

21. Draw a Fischer projection of the following molecule.

$$CH_3CH_2 - \overset{OH}{\underset{H}{C}} - CH_3$$
(with CH₃ on a wedge)

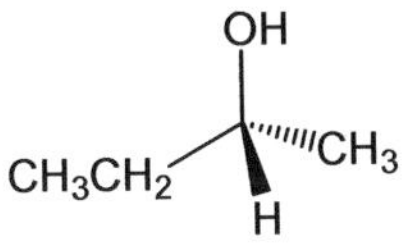

22. Draw Perspective Drawing and Fischer projection of the enantiomer of the above molecule.

23. Which of the following molecules are chiral and how many chirality centers are present in each of the chiral molecules?

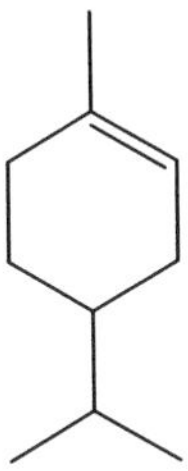

      1 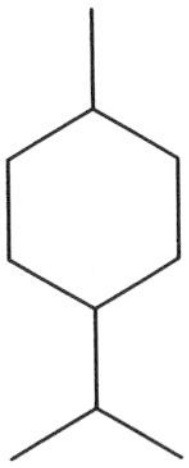   2 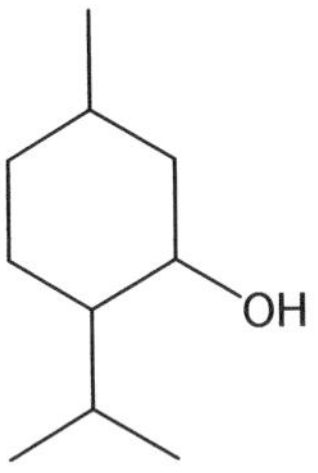   3

24. Determine configuration of the following compounds (show your work):

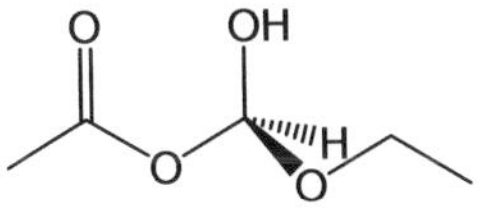

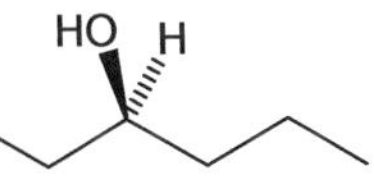

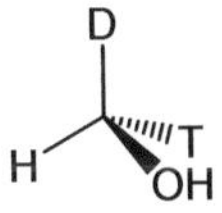

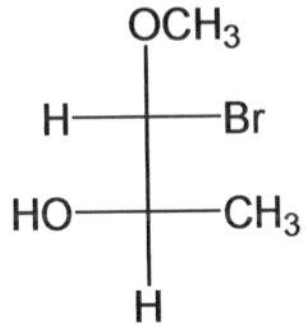

[Structures shown with stereochemistry:
- An acetate ester with OH and stereocenter
- HO, H on a carbon chain (stereocenter)
- A carbon with D, H, T, OH substituents
- A Fischer-like projection with OCH₃, Br, CH₃, HO, H, H]

25. Complete the following table.

(S)-2-bromobutane	
(R)-2-butanol	
(R)-2-amino-1-pentanol	
	(structure with OH on wedge, 4-methyl-2-pentanol type)
	(structure with Br on wedge, 3,3-dimethyl-2-butyl bromide type)

26. Draw the erythro and the threo pairs of enatiomers of 2-chloro-3-fluorobutane.

27. Provide **Fischer projection formulas** of threo isomers of 2,3-dibromobutane.

28. Label each stereocenters (chirality centers) with its relative stereochemical configuration ((R) or (S)).

29. A compound isolated as a product of a chemical reaction has $[\alpha]_D = +5.8°$. Which of the two formulas below is the correct the formula for that compound?

Questions 30 – 39: For the following pairs of compounds indicate if they are a) enantiomers, b) diastereomers c) two identical molecules of a meso compound, or d) two identical molecules but not a meso isomer.

30.

31.

32.

33.

34.

35.

36.

37.

38.

39.

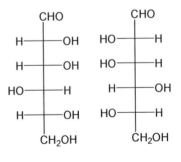

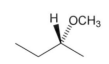

40. The relative configuration of the following compound is:

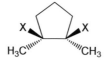

a) always (R,S)-
b) always (R,R)-
c) (R,S)- if X has higher priority than –CH$_3$.
d) (R,R)- if X has higher priority than –CH$_3$.

41. Which compound is an isomer of CH$_3$CH$_2$SCH$_2$CH$_3$?

a) CH$_3$CH$_2$SCH$_3$
b) CH$_3$CH$_2$CH$_2$SH
c) CH$_3$CH(OH)CH$_2$CH$_3$
d) CH$_3$CH(SH)CH$_2$CH$_3$

42. Which of the following compounds does not have an isomer?

a) CH$_3$-CHCl-CH$_2$-CH$_3$
b) Cl-CH$_2$=CH-Cl
c) CH$_2$=CH-Cl
d) Cl-CH$_2$-CH$_2$-Cl

43. Which of the following statements is not an essential feature of an optically active molecule? An optically active molecule:

a) rotates the plane of polarized light.
b) has a non-superimposable mirror image.
c) has no element of symmetry.
d) has an asymmetric carbon atom.

44. Which of the following molecules is chiral?

a) 1-hexanol
b) 4-octanol
c) 4-heptanol
d) 3-pentanol

45. Which of the following molecules *is not* chiral?

a) 2-methylnonane
b) 3-methylnonane
c) 4-methylnonane
d) 5-methylnonane

46. According to the Canh-Inglold-Prelog rules, which of the following is the correct priority sequence?

   a) $-CH=CH_2 > -COOH > -CHO > -CH_2CH_3$
   b) $-COOH > -CHO > -CH=CH_2 > -CH_2CH_3$
   c) $-CH=CH_2 > -COOH > -CHO > -CH_2CH_3$
   d) $-COOH > -CH_2CH_3 > -CHO > -CH=CH_2$

47. Which compounds are a pair of enantiomers?

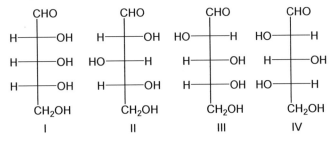

   a) I and III
   b) II and III
   c) I and IV
   d) II and IV

48. A chiral compound has $[\alpha]_D = +16°$. Its enantiomer has $[\alpha]_D$:

   a) of exactly $+16°$.
   b) of more than $+16°$.
   c) of less than $+16°$.
   d) of approximately $+16°$.

49. A sample of (R)-(+)-glyceraldehyde has optical purity of 0.60. How much (S)-(-)-glyceraldehyde is present in the sample?

   a) 20%
   b) 40%
   c) 60%
   d) 80%

50. How many stereoisomers of the following molecule are possible?

$$HOOCCH=C=CHCOOH$$

   a) two geometrical isomers
   b) two optical isomers
   c) two optical and two geometrical isomers
   d) none

Answers to Additional Exercises: Questions 14-50

14. The other conformation is:

15. Chirality centers:

16. There are three chirality centers in 2-bromonorbornane:

17. Chirality centers:

18. The two enantiomers of 2-iodobutane:

saw-horse projections

perspective line formulas

19. There are three isomers of 3,4-dichlorohexane, a pair of enantiomers and a meso compound:

```
    CH2CH3           CH2CH3           CH2CH3
H ──┼── Cl       Cl ──┼── H        H ──┼── Cl
Cl ──┼── H        H ──┼── Cl       H ──┼── Cl
    CH2CH3           CH2CH3           CH2CH3
```

20. There are three chirality centers and no plane of symmetry. Therefore, there are 8 isomers and four pairs of enantiomers.

21. Fischer projection:

```
      CH3
HO ──┼── H
     CH2CH3
```

22. Enantiomer:

```
      CH3
HO ──┼── H
     CH2CH3
```

23. The molecule **1** has one chirality center. It is the carbon 4. Molecule **2** has no chirality centers. Conversion of the double bond into a single bond made two "halves" of the ring identical and carbon 4 is no longer chiral. Introduction of hydroxyl group into molecule **2** gives the molecule **3**. Two alkyl bearing carbon atoms are now both chirality centers. In addition, carbon bonded to –OH is also chiral. Therefore, molecule **3** has three chirality centers.

24. Relative configuration:

97

25. Table:

(S)-2-bromobutane	[structure: 2-bromobutane with Br on dashed wedge]
(R)-2-butanol	[structure: 2-butanol with OH]
(R)-2-amino-1-pentanol	[structure: HO–CH₂–C(NH₂)(H)–propyl]
(R)-5-methyl-3-hexanol	[structure with OH]
(R)-3-bromo-2,2-dimethylbutane	[structure with Br]

26. Erythro and the threo pairs of enatiomers of 2-chloro-3-fluorobutane:

[Fischer projections: erythro pair and threo pair]

erythro pair   threo pair

27. Fischer projection formulas of threo isomers of 2,3-dibromobutane:

[Two Fischer projections]

28. Stereocenters:

[Cyclohexane structure with Cl (S), CH₃ (S), Br (S) substituents]

29. Only the compound A is a chiral compound, compound B is not. Therefore, the correct answer is compound A.
30. d); 31. a); 32. d); 33. b); 34. a); 35. a); 36. c); 37. a); 38. a); 39. a); 40. a) this is a meso compound, which is always (R,S); 41. d); 42. c); 43. c); 44. b); 45. c); 46. b); 47. d); 48. a); 49. a); 50. b).

Exercises in Organic Chemistry

Special Topic: Stereochemistry in Organic Reactions

Topics to review: stereochemical outcome of a reaction, stereoselectivity and stereoselective reactions of alkanes, alkenes and alkynes.

Questions 1-6.

1. Provide Fischer Projection formula(s) for the monochlorination product(s) of the following reaction. Assume that the chlorination occurs on the carbon 3 of (S)-2-chlorobutane.

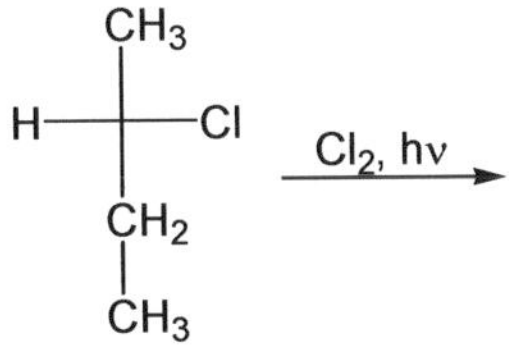

2. Provide a mechanism for the previous reaction. Explain formation of the product(s).

3. Provide formula of the product of each reaction shown below. Are the products chiral?

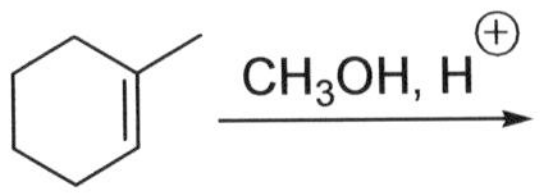

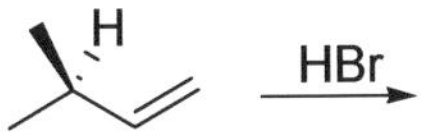

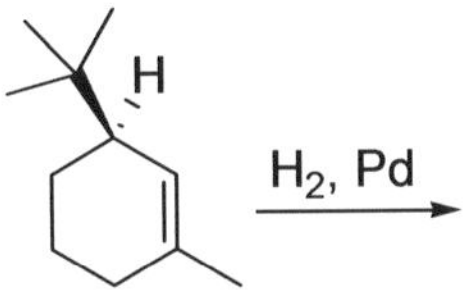

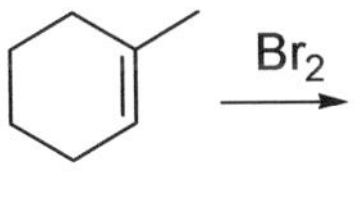

4. Draw the product(s) of hydroboration-oxidation of (E)-3,4-dimethyl-3-hexene. Is the reaction stereoselective? Is the reaction stereospecific?

5. In which of the following reactions may the mode of addition ("*syn*" – from the same side or "*anti*" from the opposite sides) be deduced from the stereochemistry of the product. Addition of:

    a) HBr to propene
    b) HBr to cyclopentene
    c) HBr to *trans*-2-butene
    d) HBr to 1,2-dimethylcyclohexene

6. An optically active alcohol, A, has the molecular formula $C_6H_{12}O$ and contains -CH=CH$_2$ group and at least one -CH$_3$ group. When the compound A is reduced with hydrogen over nickel the reaction product is optically inactive. What is the structure of A?

a)   b)   c)   d)

Answers to questions 1-6.

1. Reaction products are the two diastereomers. Chirality center on the carbon 2 remains unchanged.

2. The intermediate is a free radical. Chlorine molecule reacts with the intermediate radical by approaching it from either side of the unhybridized *p* orbital. Although, due to the presence of a neighboring chirality center, approach for one side is preferred. Due to the steric hindrance, in the drawing below the bottom side approach appears to be favored to the top side approach. Thus, both diastereomers are obtained but in different amounts.

3. Reaction products:

a single isomer is the reaction product

single enantiomer + a mixture of diastereomers is obtained by attack of Br⁻ on the intermediate carbocaiton.

single enantiomer
One enantiomer is the reaction product. Hydrogenation occurs from the less hindered side opposite to the bulky *t*-butyl substituent.

Product is a mixture of two enantiomers as a result of approach of Br⁻ from the side opposite to the intermediate bromonium ion.

4. Hydroboration-oxidation sequence results in an addition of –H and –OH from the same side of a double bond (*syn* addition). The reaction product is threo pair of enantiomers. Therefore, the reaction is stereoselective since one pair of enantiomers is produced out of the possible two. It is also stereospecific since one stereoisomer of the starting material (*E*) produces one pair of enantiomers (threo) and the other one (*Z*) produces the other pair of enantiomers (erythro). The best way to represent the reaction outcome is by means of Fischer projections.

5. One should consider the products of each reaction and what their structure tell us about the mode of addition.

Reaction a) does not give any stereoisomers and can be easily eliminated. Reaction b) is not a correct answer either. The product is a mixture of enantiomers. Some students select this answer because the starting material is one stereoisomer. However, addition in either *syn* or *anti* mode gives the same product. Reaction c) is an addition to a cyclic substrate. However, it is not a correct answer either. Again, addition in either mode gives the same product. Finally, the answer d) is the correct one. In this case, *syn* addition gives *cis* isomer and anti addition the *trans* isomer.

One should keep in mind that addition of HBr in general is not a stereoselective reaction and an addition to small and medium ring cycloalkenes is an exception.

6. Again one should consider the reaction products. This time one should examine the products of hydrogenation of each of the four alcohols. Even after the double bond has been hydrogenated, alcohols a) and d) are still chiral. Therefore, we can eliminate those. Among the two remaining answers, only the alcohol c) has a methyl group and is the correct answer.

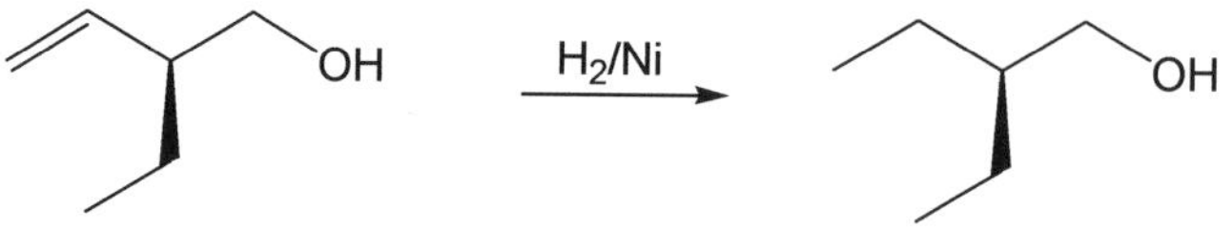

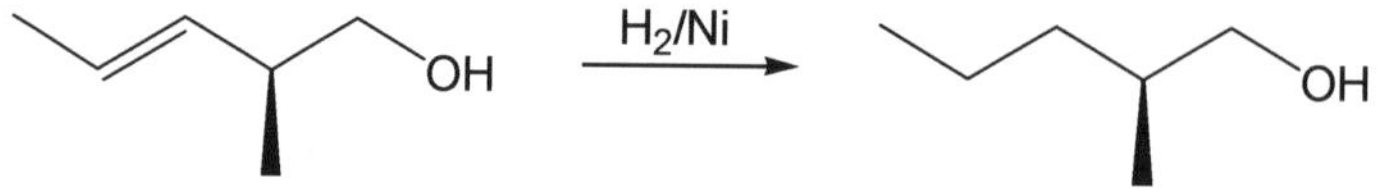

Additional Exercises: Questions 7-11

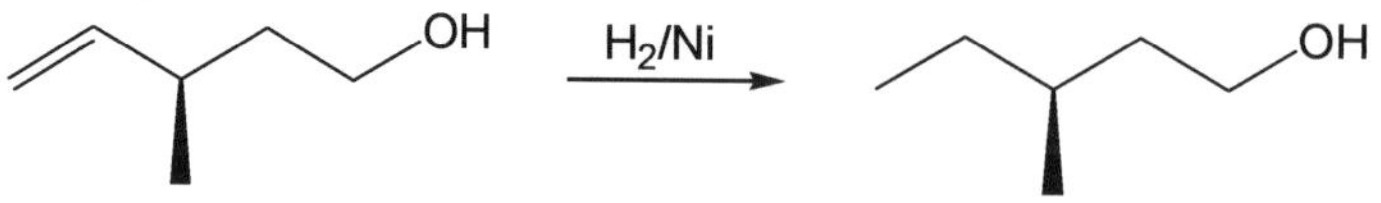

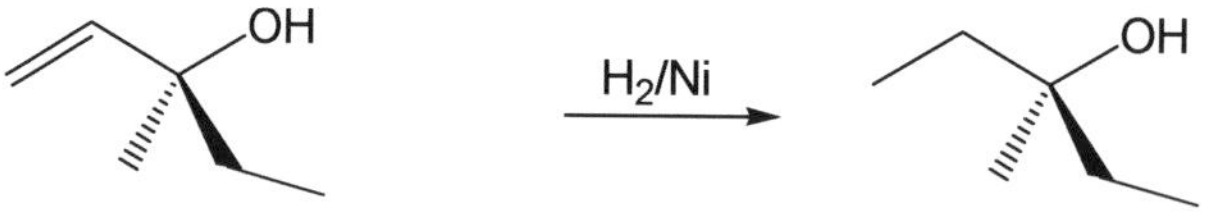

7. Provide **Fischer Projection formula(s)** for the monobromination product(s) of the following reaction. Assume that bromination occurs on the carbon 3.

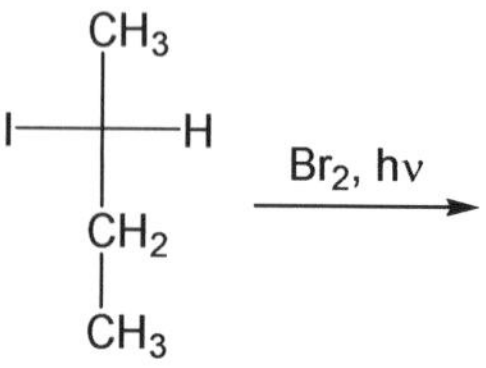

Exercises in Organic Chemistry

8. Provide formulas of the major products of each of the following reactions including correct stereochemistry.

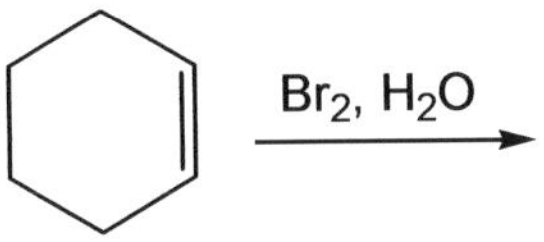

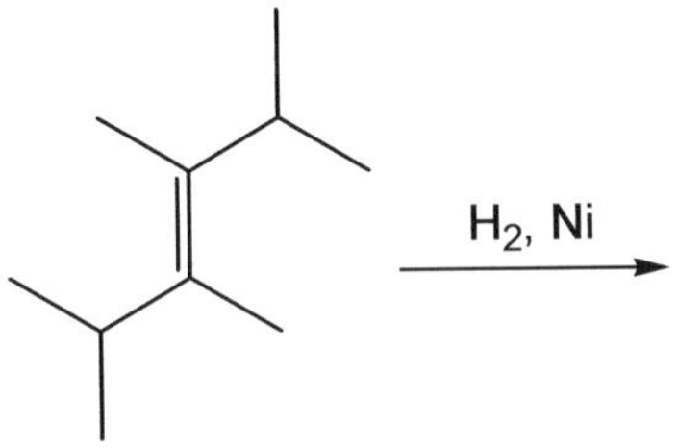

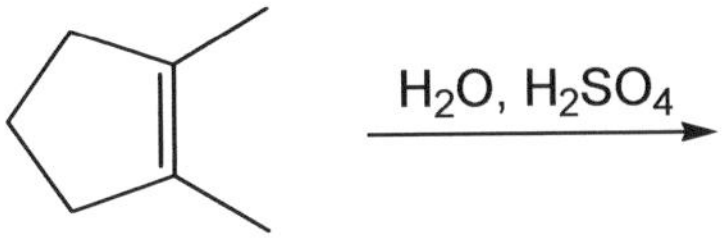

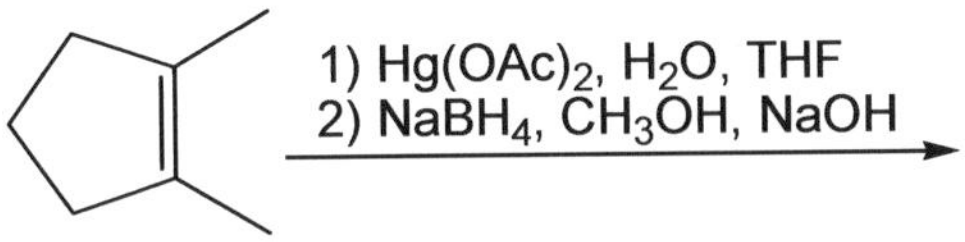

   R—C≡C—R' $\xrightarrow{\text{Br}_2, \text{NaCl}}$

   R—C≡C—R' $\xrightarrow{\text{1) Cl}_2 \text{ 2) Br}_2}$

   (cyclohexene) $\xrightarrow{\text{Br}_2, \text{H}_2\text{O}}$

   (2,4-dimethyl-3-isopropyl-alkene structure) $\xrightarrow{\text{H}_2, \text{Ni}}$

   (1,2-dimethylcyclopentene) $\xrightarrow{\text{H}_2\text{O}, \text{H}_2\text{SO}_4}$

   (1,2-dimethylcyclopentene) $\xrightarrow{\text{1) Hg(OAc)}_2, \text{H}_2\text{O}, \text{THF} \quad \text{2) NaBH}_4, \text{CH}_3\text{OH}, \text{NaOH}}$

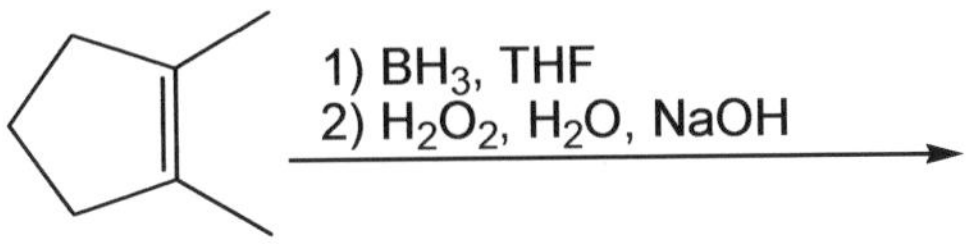

   (1,2-dimethylcyclopentene) $\xrightarrow{\text{1) BH}_3, \text{THF} \quad \text{2) H}_2\text{O}_2, \text{H}_2\text{O}, \text{NaOH}}$

9. Provide formulas of the starting materials including correct stereoisomer (where applicable).

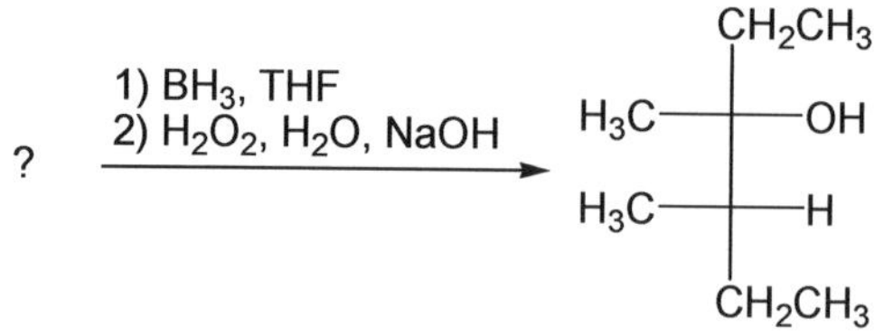

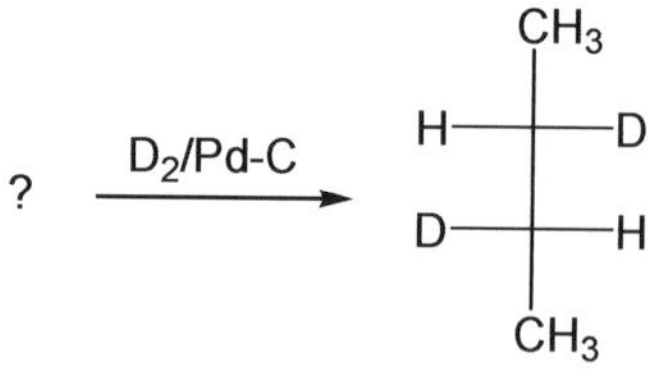

? → 1) BH₃, THF 2) H₂O₂, H₂O, NaOH → Fischer projection: CH₂CH₃ / H₃C—OH / H₃C—H / CH₂CH₃

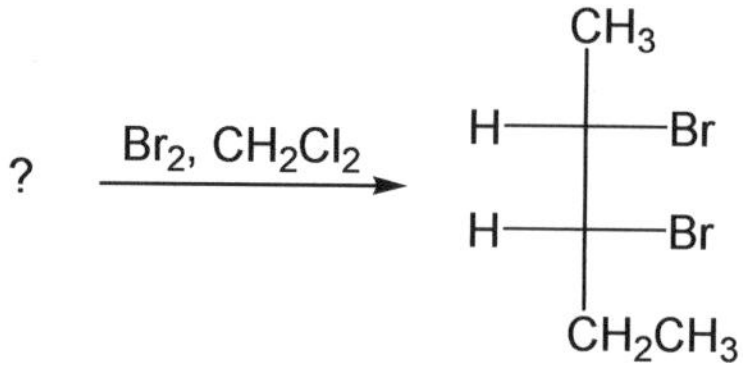

? → Br₂, CH₂Cl₂ → Fischer projection: CH₃ / H—Br / H—Br / CH₂CH₃

? → D₂/Pd-C → Fischer projection: CH₃ / H—D / D—H / CH₃

10. Provide a reagent for each of the following reactions.

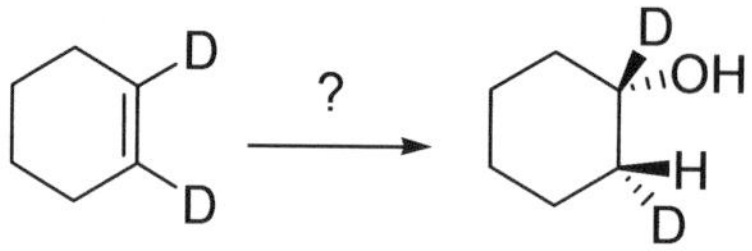

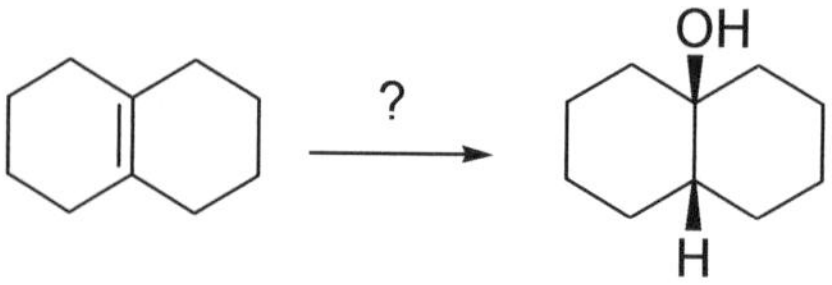

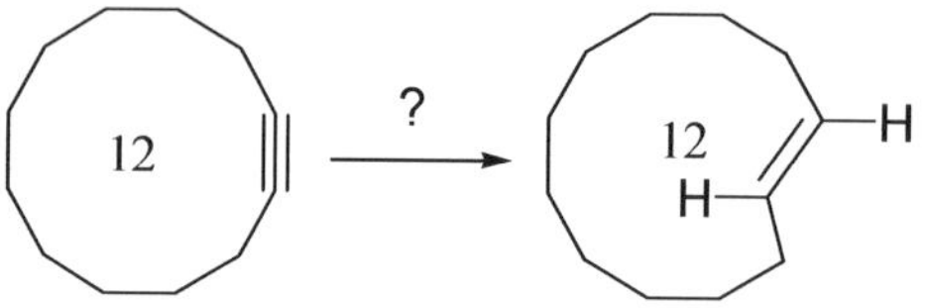

cyclohexene with D,D on double bond → ? → cyclohexane with D, OH (wedge), H, D (dash)

bicyclic alkene (octahydronaphthalene-type) → ? → trans-decalin with OH (wedge) and H (dash)

cyclododecyne (12) → ? → cis-cyclododecene (12) with H, H on same side

11. In which of the following addition reactions may the mode of addition ("*syn*" – from the same side or "*anti*" from the opposite sides) of bromine to an alkene be deduced from the stereochemistry of the product:

a) methylenecyclohexene (C₆H₁₀=CH₂)
b) 2,3-dimethyl-2-butene
c) 1-butene
d) *trans* 2-butene

Answers to Additional Exercises: Questions 7-11

7. Bromination of 2-iodobutane:

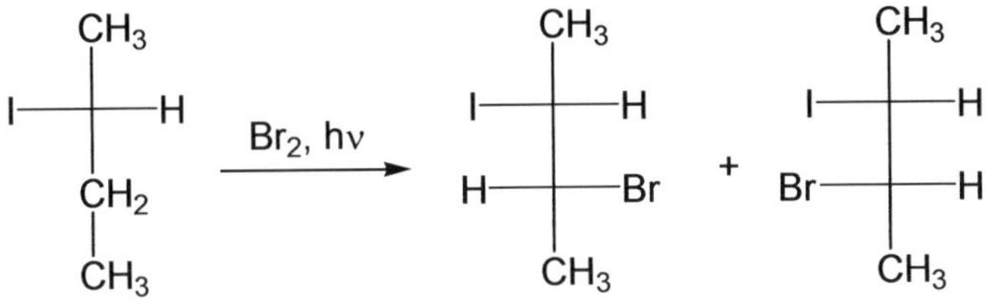

[Structure: 2-iodobutane with CH₃, I, H, CH₂, CH₃ → Br₂, hν → two Fischer-projection products with I, CH₃ at top and CH₃ at bottom; one has H–Br (left H, right Br) and the other has Br–H (left Br, right H)]

8. Reaction products:

- Alkyne + Br₂, NaCl → 2-bromo-3-chloro-2-butene (Br and Cl on double bond with two CH₃)
- Alkyne: 1) Cl₂ 2) Br₂ → tetrahalide with Cl, Br, Br, Cl (or shown as Fischer with CH₃, Cl–Br, Cl–Br, CH₃)
- Cyclohexene + Br₂, H₂O → trans-2-bromocyclohexanol (Br and OH trans)
- Tetrasubstituted alkene + H₂, Ni → alkane (H, CH₃, H, CH₃ with isopropyl groups)
- 1,2-dimethylcyclopentene + H₂O, H₂SO₄ → 1,2-dimethylcyclopentanol (OH and H, Markovnikov)
- 1,2-dimethylcyclopentene: 1) Hg(OAc)₂, H₂O, THF 2) NaBH₄, CH₃OH, NaOH → 1,2-dimethylcyclopentanol (anti-Markovnikov stereochem shown)
- 1,2-dimethylcyclopentene: 1) BH₃, THF 2) H₂O₂, H₂O, NaOH → trans-1,2-dimethyl-2-hydroxycyclopentane (OH and H syn addition)

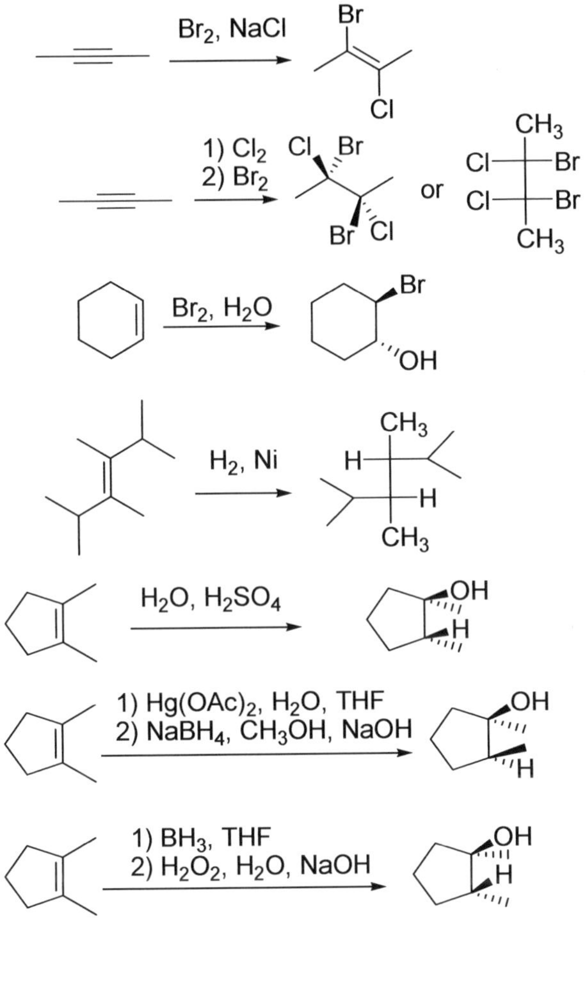

9. Starting materials:

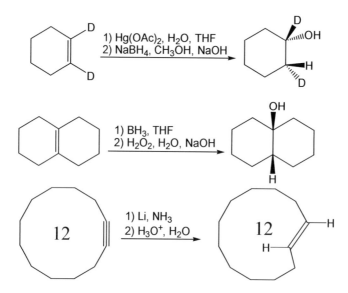

10. Reagents:

11. d).

Exercises in Organic Chemistry

12. Substitution Reactions of Alkyl Halides

Topics to review: substitution reactions of alkyl halides, S_N1 and S_N2 reaction mechanisms, steric and electronic effects, strength of nucleophiles, leaving groups, effect of the solvent, stereochemical outcome of a reaction, rearrangements.

Questions 1-4.

1. How would you determine whether the following substitution reaction proceeds by S_N1 or S_N2 mechanism?

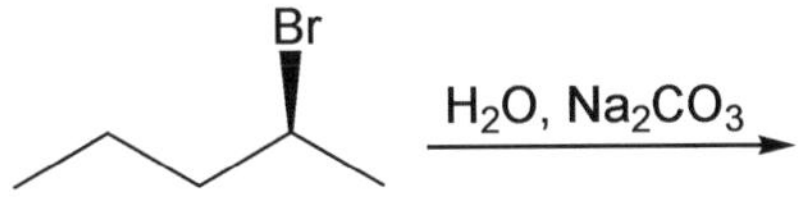

2. Explain by means of chemical equations how would you prepare (S)-2-iodopentane from (R)-2-chloropentane.

3. Provide formulas of the two starting alkyl halides, R-X (not taking into account that X may be different halogen atoms – Cl, Br or I), that upon reaction with water would produce 2,3-dimethyl-2-butanol.

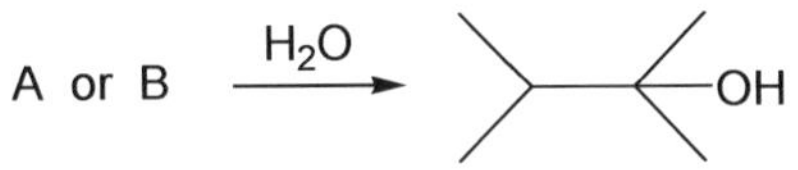

4. Which of the following reactions is going to be faster?

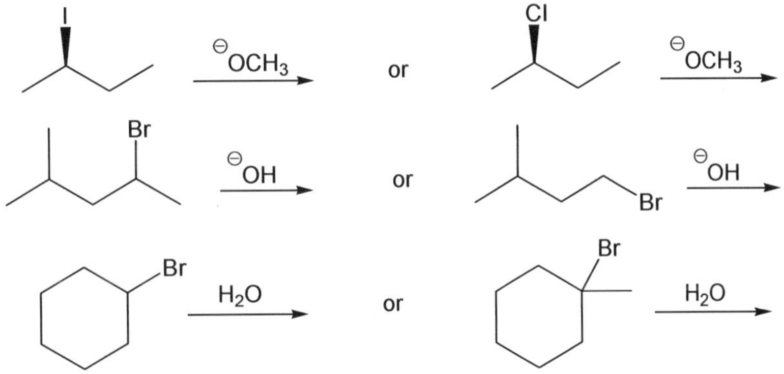

Answers to questions 1-4.

1. The starting material is (S)-2-bromopentane – it is a single enantiomer. An S_N1 reaction proceeds with a complete or partial racemization, while an S_N2 reaction proceeds with an inversion of configuration. Therefore, at the end of reaction, one should isolate the product, 2-pentanol, and examine its optical activity, if any. Pure (R)-2-pentanol would indicate that the mechanism is S_N2, while a mixture of (R)- and (S)- isomers would indicate S_N1 mechanism.

2. An S_N2 reaction proceeds with an inversion of configuration. Therefore, one should convert (R)-2-chloropentane to (S)-2-iodopentane in an S_N2 reaction. Alkyl chlorides and bromides are converted into the corresponding alkyl iodides by treatment with sodium iodide in acetone:

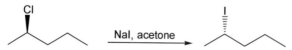

3. The two halides are 2-halo-2,3-dimethylbutane and 2-halo-3,3-dimethylbutane. The latter gives the reaction product after a rearrangement.

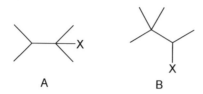

4. Reaction rates:

![reaction 1]

This is a faster reaction
- iodine is a better leaving group.

![reaction 2]

This is a faster reaction
- a primary substrate reacts faster in an S_N2 reaction.

![reaction 3]

This is a faster reaction
- a tertiary substrate reacts faster in an S_N1 reaction.

Exercises in Organic Chemistry

Question 5-10.

5. What are the starting materials for the following reactions?

   ? ⟶ PhO⁻Na⁺ + cyclohexyl-CH₂-O-Ph

   ? + ⁻OH ⟶ trans-1-methyl-2-hydroxycyclohexane (OH and CH₃ trans)

   ? + (CH₃)₂C=C:⁻ Na⁺ ⟶ (CH₃)₂CH−C≡C−CH(CH₃)₂ (approx: 2,5-dimethyl-3-hexyne)

   ? + CH₃OH ⟶ compound with OCH₃ and OCH₂CH₃ on adjacent carbons

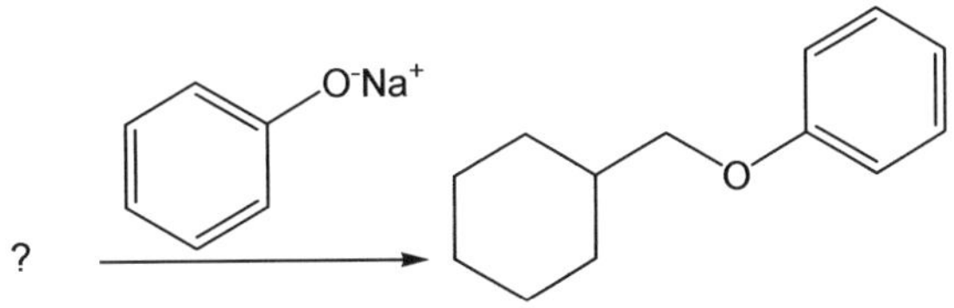

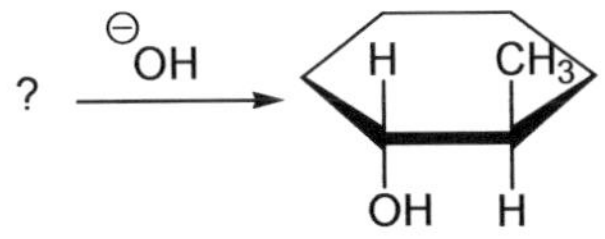

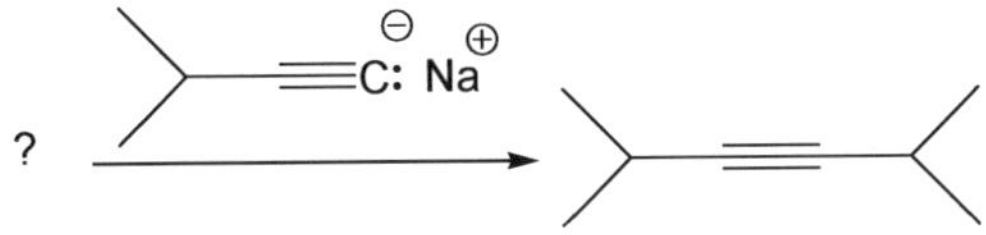

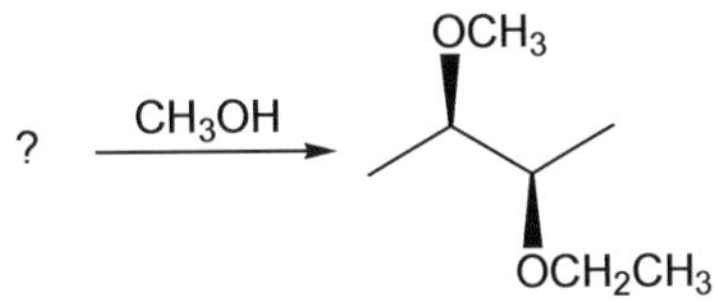

6. What is the stereochemical outcome of the following reaction?

   (sec-butyl chloride) —NaI/acetone→ (sec-butyl iodide)

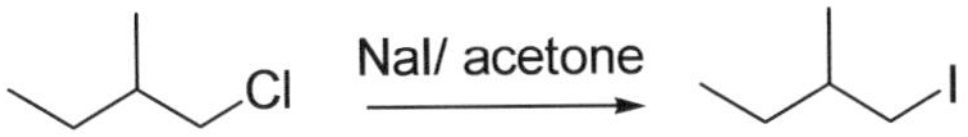

   a) The reaction proceeds with retention of configuration.
   b) The reaction proceeds with inversion of configuration.
   c) The reaction proceeds with racemization.
   d) One cannot tell from the example above.

7. What are the products of the following reactions?

   CH₃−CH₂−I + (CH₃)₃N ⟶

   2-bromopentane + NaI, acetone ⟶

   PhCH₂−I + H₂O ⟶

   (Br)(Br)C=CH−CH(CH₃)− + CH₃S⁻Na⁺ ⟶

   trans-1-iodo-2-?-cyclopentane + ⁻O−Na⁺ ⟶

   (neopentyl-type iodide) + CH₃OH ⟶

109

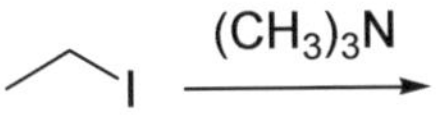

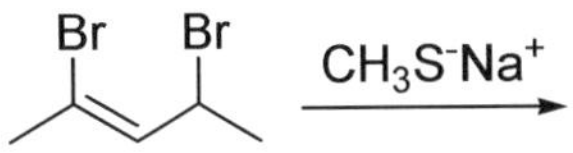

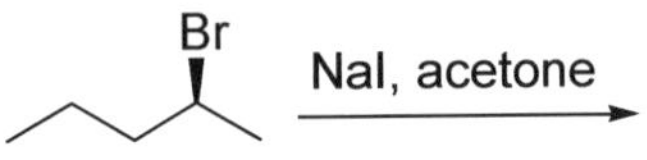

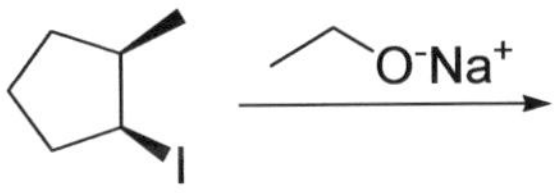

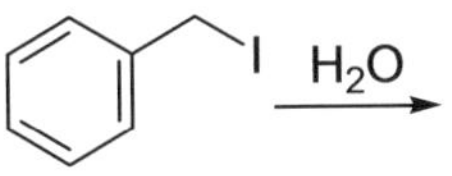

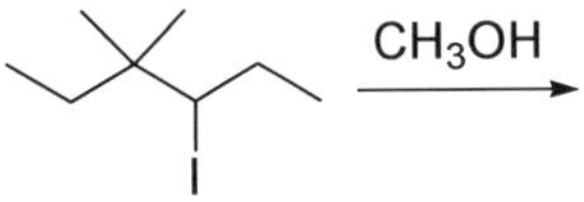

8. What are the reagents for the following reactions?

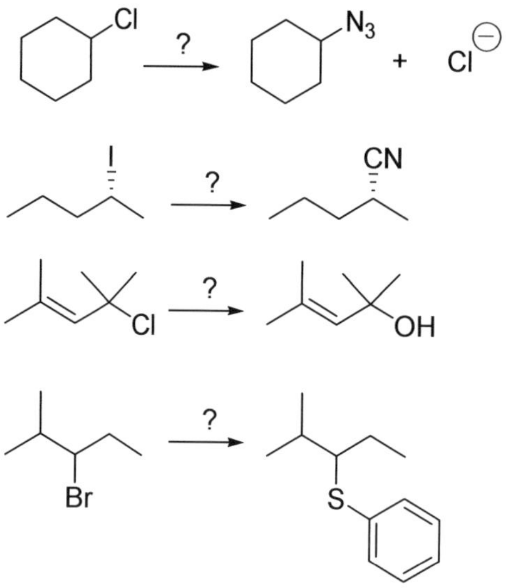

   Cyclohexyl-Cl → Cyclohexyl-N₃ + Cl⁻

   (sec-pentyl-I) → (sec-pentyl-CN)

   (3-methyl-2-butenyl chloride) → (3-methyl-2-buten-2-ol, tertiary OH)

   (2-bromo-3-methylpentane) → (3-(phenylthio)-2-methylpentane)

9. Why these reactions do not work?

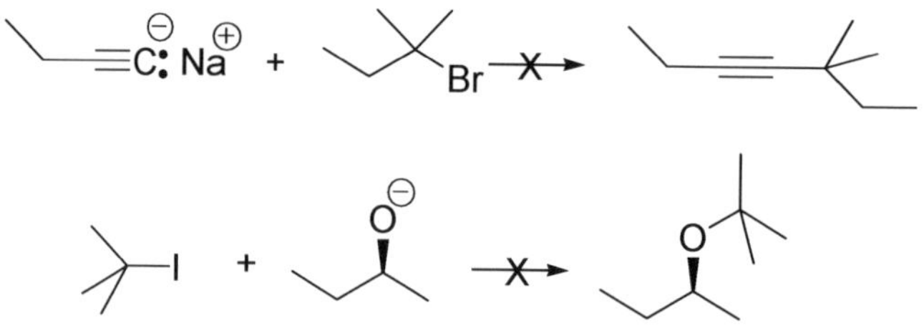

   HC≡C:⁻ Na⁺ + (t-butyl bromide) —X→ (alkyne product)

   (t-butyl iodide) + CH₃C(=O)O⁻ —X→ (t-butyl isopropyl ester)

10. Indicate the reaction mechanism in the space provided.

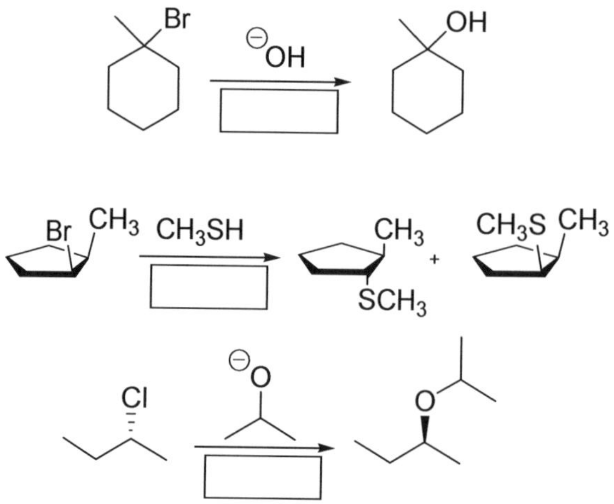

    (1-bromo-1-methylcyclohexane) + ⁻OH → (1-methylcyclohexanol)

    (1-bromo-2-methylcyclopentane) + CH₃SH → (1-methyl-2-(methylthio)cyclopentane) + (1-(methylthio)-2-methylcyclopentane)

    (sec-butyl chloride) + CH₃C(=O)O⁻ → (sec-butyl isopropyl ether)

110

11. Formula of 9-bromotripticene is shown on the right. How would you prepare this molecule? Can the molecule undergo S_N2 reactions? Would you expect it to undergo S_N1 reactions? Consider both kinetic and thermodynamic stability. How stable would you expect the corresponding carbocation to be? Is it stabilized by resonance with the three benzene rings? Why, or why not?

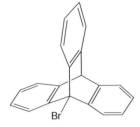

Answers to question 5-11.

5. Starting materials:

6. d) One cannot tell. The starting material is not a chiral compound. Therefore, we cannot make a conclusion about the stereochemical outcome of a reaction.

7. Reaction products:

CH₃CH₂I + (CH₃)₃N → CH₃CH₂N⁺(CH₃)₃ I⁻

(S)-2-bromopentane + NaI, acetone → (R)-2-iodopentane

PhCH₂I + H₂O → PhCH₂OH

(E)-2,3-dibromo-2-butene + CH₃S⁻Na⁺ → (E)-2-bromo-3-(methylthio)-2-butene

Allylic bromine reacts, vinylic does not.

trans-2-methyl-1-iodocyclopentane + CH₃CH₂O⁻Na⁺ → trans-2-methyl-1-ethoxycyclopentane

neopentyl-type iodide + CH₃OH → rearranged methyl ether (OCH₃)

Reaction proceeds with a rearrangement.

8. Reagents:

cyclohexyl chloride + Na⁺N₃⁻ → cyclohexyl azide + Cl⁻

(S)-2-iodopentane + K⁺CN⁻ → (R)-2-cyanopentane

3-chloro-2-methyl-2-butene-type + Na⁺OH⁻ → corresponding allylic alcohol

2-methyl-3-bromopentane + PhS⁻Na⁺ → 3-(phenylthio)-2-methylpentane

9. Those are S_N2 reactions and S_N2 reactions do not work on tertiary substrates due to steric hindrance.

10. Reaction mechanisms:

Reaction on a tertiary haloalkane.

Reaction proceeds with racemization.

Reaction proceeds with inversion of configuration.

11. 9-Bromotripticene can be prepared in a free radical substitution reaction between bromine and tripticene. It is a tertiary alkyl halide and cannot undergo S_N2 reactions. In addition, a bicyclic system prevents any nucleophile form approaching the bromine-bearing carbon from the opposite side. The molecule is capable forming a tertiary carbocation and undergoes S_N1 reactions. The carbocation has a stability of an ordinary tertiary carbocation. In fact, the stability may be a somewhat lower due to angle strain it introduces into the rings. Carbocation has a considerable kinetic stability as the arrangement of the three benzene rings makes approach of a nucleophile difficult. It is not stabilized by resonance with the three benzene rings because the three rings are fixed in positions that prevent overlap of their π orbitals with the carbocation's empty p orbital.

Additional Exercises: Questions 12-22

12. Predict the major product for each of the following reactions. Indicate stereochemistry of the product when appropriate.

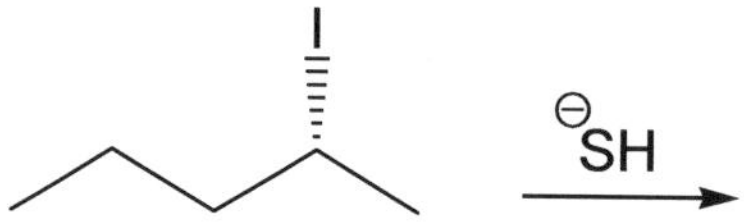

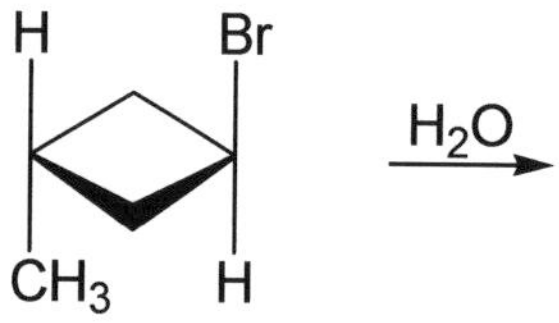

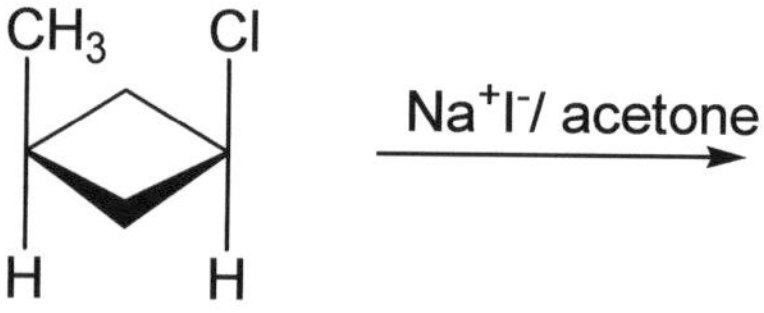

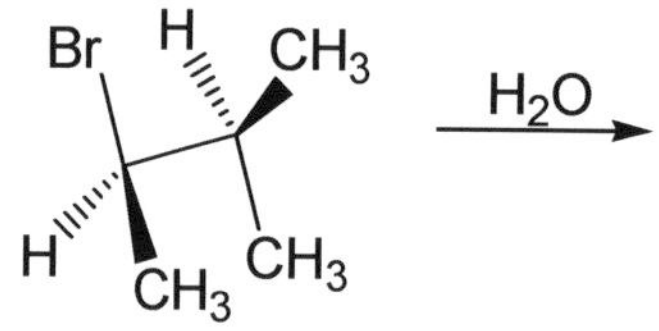

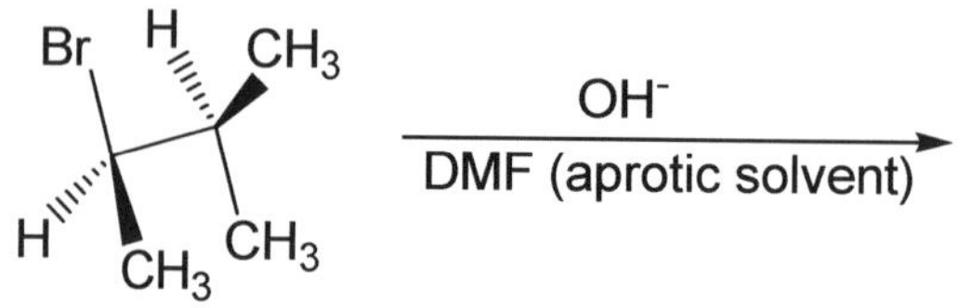

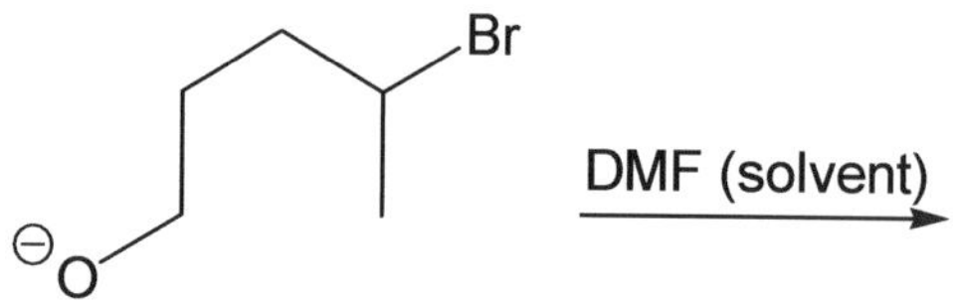

Exercises in Organic Chemistry

13. Provide formulas of the reagents for the following reactions.

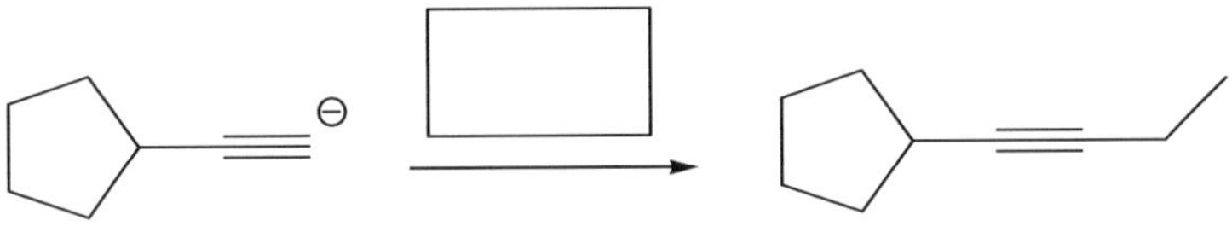

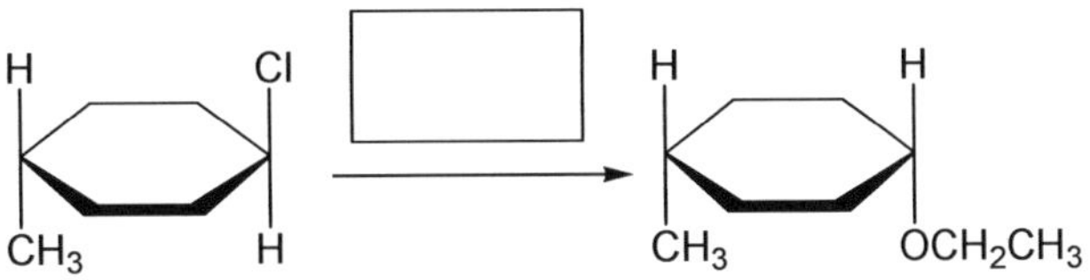

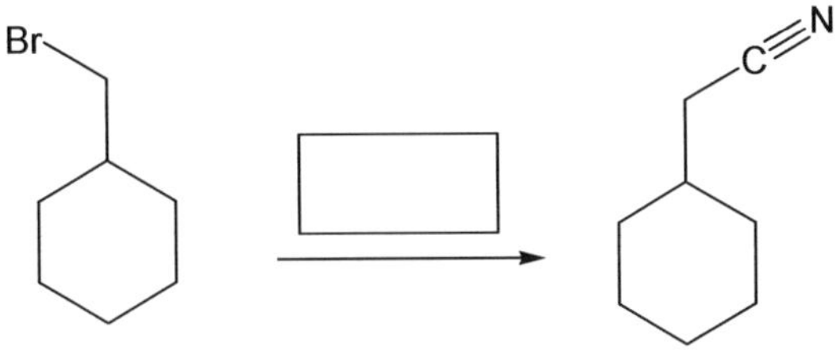

14. Provide formulas of the starting materials for the following reactions.

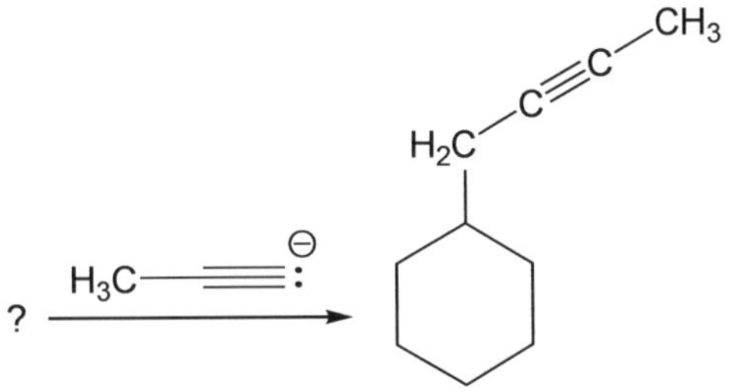

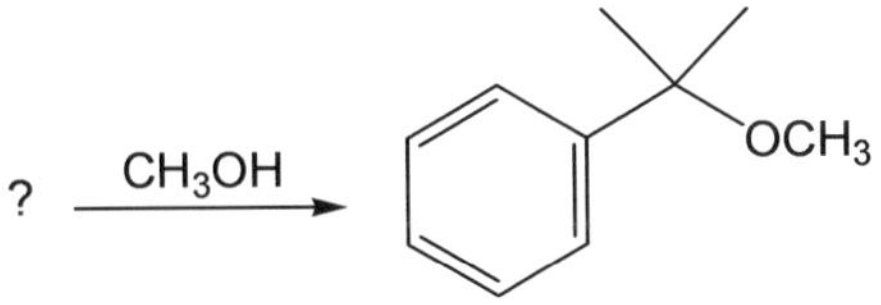

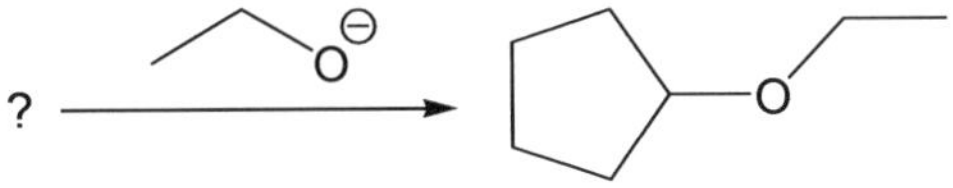

15. For the following reactions provide the formulas of the missing compounds:

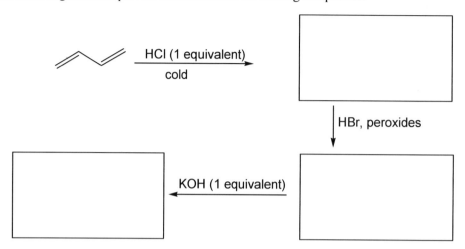

16. Which of the following is the strongest nucleophile?

    a) CH_3OH
    b) CH_3O^-
    c) HO^-
    d) $(CH_3)_3CO^-$

17. Which of the following nucleophiles will favor S_N2 reaction mechanism rather than S_N1?

    a) CH_3O^-
    b) H_2O
    c) $(CH_3)_3COH$
    d) NH_3

18. Which of the following nucleophiles will favor S_N1 reaction mechanism rather than S_N2?

    a) OH^-
    b) CH_3O^-
    c) $(CH_3)_3COH$
    d) CH_3OH

19. The complete racemization which occurs during the hydrolysis of a haloalkane suggests that the reaction is:

    a) S_N1
    b) S_N2
    c) Free radical substitution
    d) It is not possible to determine. One has to check for optical activity of the product.

Exercises in Organic Chemistry

20. Which of the following reagents reacts with 2-bromopropane in a substitution reaction?

    a) HCl
    b) C₂H₅OCH₃
    c) CH₃COOH
    d) KCN

21. The change in hybridization of carbon in the course of an S_N2 reaction is:

    a) sp^3 to sp^2
    b) sp^3 to sp
    c) sp to sp^3
    d) sp^2 to sp

22. Which of the following compounds reacts the fastest with water?

    a) *t*-butyl chloide
    b) 1-chloro-1-butene
    c) 1-chlorobutane
    d) 2-chlorobutane

Answers to Additional Exercises: Questions 12-22

12. Reaction products:

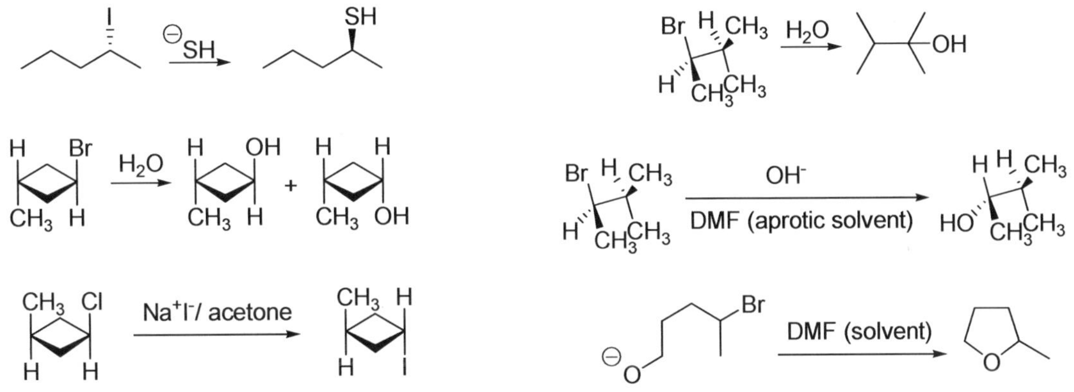

13. Reagents:

(cyclopentyl-C≡C:⁻) + CH₃CH₂X (X = Cl, Br or I) → cyclopentyl-C≡C-CH₂CH₃

trans-1-chloro-2-methylcyclohexane + CH₃CH₂O⁻ → trans-1-ethoxy-2-methylcyclohexane

(bromomethyl)cyclohexane + ⁻C≡N → (cyanomethyl)cyclohexane

14. Starting materials:

cyclohexyl-CH₂-X + H₃C-C≡C:⁻ → cyclohexyl-CH₂-C≡C-CH₃

PhC(CH₃)₂-X + CH₃OH → PhC(CH₃)₂-OCH₃

cyclopentyl-X + CH₃CH₂O⁻ → cyclopentyl-OCH₂CH₃

X = Cl, Br or I

15. Missing compounds:

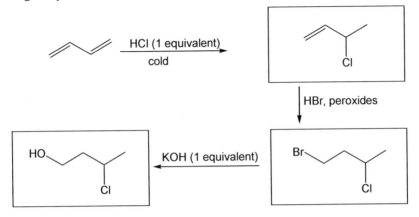

16. c); 17. a); 18. d); 19. a); 20. d); 21. a); 22. a).

Veljko Dragojlovic

Exercises in Organic Chemistry

Practice Test 3: CHAPTERS 9-12

1. Which compound is a structural isomer of $CH_3CH_2CH_2CH_2OH$?

   a) $CH_3CH_2OCH_3$
   b) $CH_3CH_2CH_2OH$
   c) $CH_3CH_2CH_2CH_3$
   d) $CH_3CH(OH)CH_2CH_3$

2. Which of the following statements is not an essential feature of an optically active molecule? An optically active molecule:

   a) rotates the plane of polarized light
   b) has a non-superimposable mirror image
   c) has no elements of symmetry
   d) has an asymmetric carbon atom

3. How many stereoisomers of the following molecule are possible?

   HOOCCH=C=CHCOOH

   a) two optical isomers
   b) two geometrical isomers
   c) two optical and two geometrical isomers
   d) none

4. How many *pairs of enantiomers* are possible in the following molecule?

   $CH_3CHOHCHOHCHOHCH_2OH$

   a) two
   b) four
   c) six
   d) eight

5. For the following pairs of compounds indicate if they are (i) enantiomers, (ii) diastereomers, (iii) two formulas of a meso isomer, or (iv) two formulas of the same compound, which is not a meso isomer.

   a)   b)

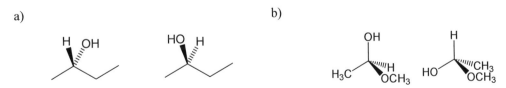

c)

```
        CH₂OH              CH₂OH
    H ──┼── OH         HO ──┼── H
   HO ──┼── H           H ──┼── OH
   HO ──┼── H           H ──┼── OH
    H ──┼── OH         HO ──┼── H
        CH₂OH              CH₂OH
```

d)

```
         CHO                CHO
    H ──┼── OH          H ──┼── OH
   HO ──┼── H          HO ──┼── H
   HO ──┼── H           H ──┼── OH
    H ──┼── OH          H ──┼── OH
        CH₂OH              CH₂OH
```

6. Determine the relative configuration of each of the following compounds (Show your work):

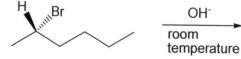

```
        CH₂OH                  CH₂OH
   HO ──┼── H              H ──┼── OH
   HO ──┼── CH₂OH          H ──┼── I
        H                  H ──┼── OH
                               CH₂I
```

7. How would you prepare the following compounds.

   a) ethyl phenyl ether

   b) [structure: CH₂=CH-CH₂-C≡CH]

   Starting with organic compounds that do not contain more than three carbon atoms.

8. Give the formula and name the product of the following reaction:

   [structure showing branched alkane with H and Br on a stereocenter] $\xrightarrow{\text{OH}^-, \text{ room temperature}}$

122

Exercises in Organic Chemistry

9. For each of the following pairs circle the better leaving group:

    a) I–   or   Cl–

    b) H₂O⁺–   or   HO–

    c) H₂N–   or   H₃N⁺–

10. Provide a mechanism and predict the product of the following reaction:

    (2-iodo-3,3-dimethylbutane) + H₂O, Ag⁺ → alcohol + AgI + H⁺

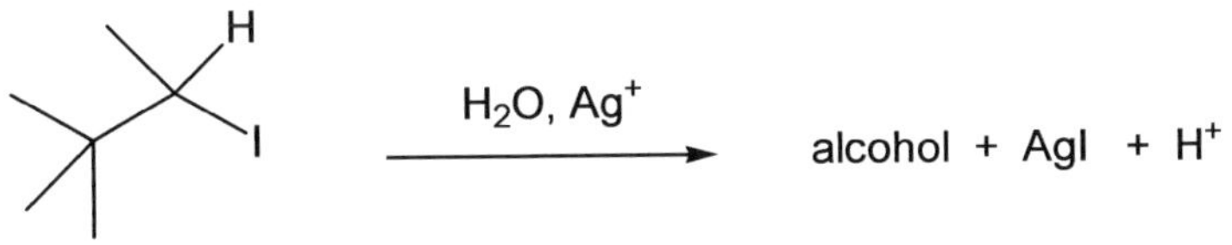

11. For the following reactions provide formulas of the missing compounds.

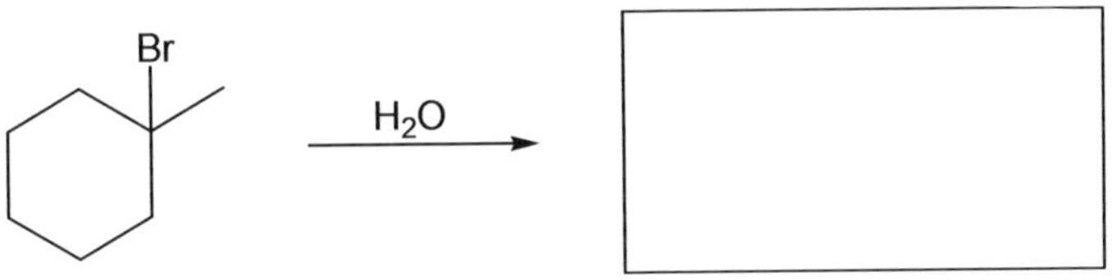

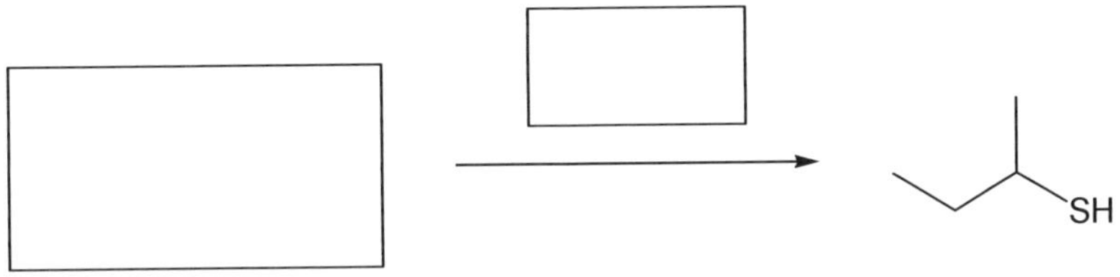

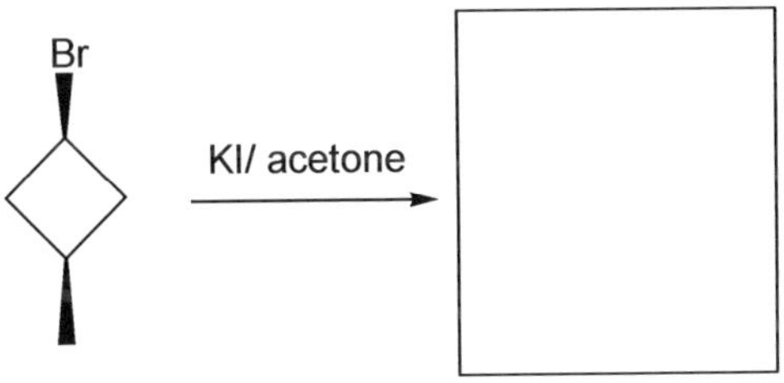

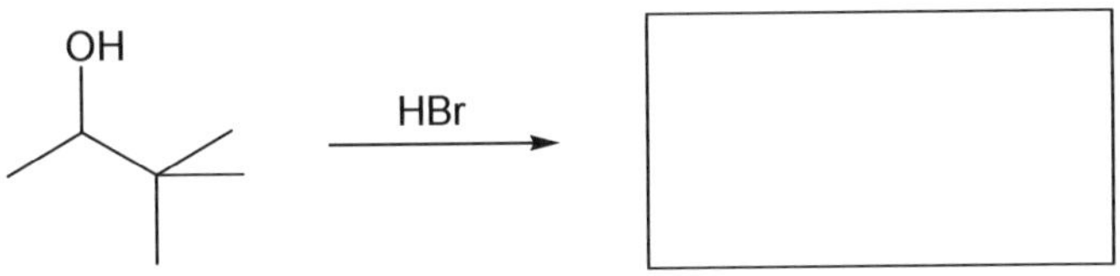

    (1-bromo-1-methylcyclohexane) + H₂O → [   ]

    [   ] → [   ] → (CH₃CH₂CH(CH₃)SH)

    (bromocyclobutane) + KI/acetone → [   ]

    (3-methyl-2-butanol, OH on C2) + HBr → [   ]

123

12. Propose a synthesis of the following compound from an alcohol, alkyl halide and any necessary inorganic reagents.

13. Provide a reasonable mechanism for the reaction given below. Include all the reagents involved, indicate electron flow by means of curved arrows, clearly label all the charges (if any), and name the mechanism.

$$H_3C-\equiv-CH_3 \xrightarrow{Na/NH_3} \text{(trans-2-butene)}$$

14. Provide a mechanism and the final product (with correct stereochemistry) for the reaction given below. Indicate electron flow by means of curved arrows, clearly label all the charges (if any).

$$\xrightarrow{Br_2 \text{ (1 equvalent)}/CH_2Cl_2}$$

15. By means of chemical equations, show how would you prepare the following compounds.

   - $CH_3CHCl_2CHCl_2CH_3$ (geminal dichloride drawn) from a compound with four carbon atoms and any necessary inorganic reagents.

   - cis-2-butene from $H_3C-C\equiv C-CH_3$

   - (Z)-2-bromo-2-butene from $H_3C-C\equiv C-CH_3$

16. Complete the following table giving the structure or name for each molecule.

	(structure: 4-methyl-2-pentyne)
3-methyl-1-pentyne	
acetylene	

Exercises in Organic Chemistry

Practice Test 3 Solutions

1. d)
2. d)
3. a)
4. b)
5. a) (i); b) (iv); c) (iii); e) (ii).
6. Relative configurations:

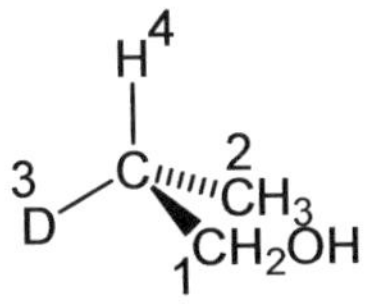

(R) 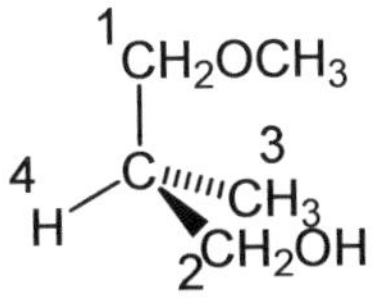   (S) 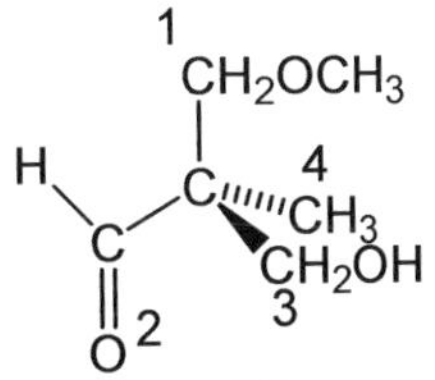   (S)

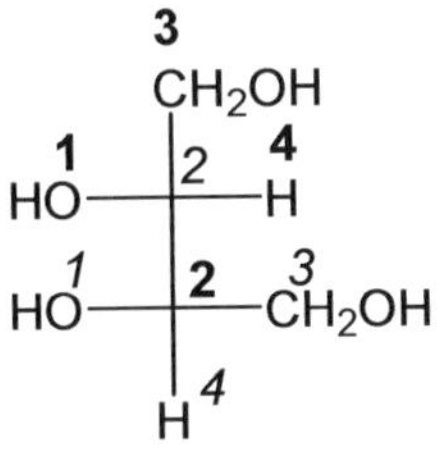

(2R,3R)   (2S,3S,4R)

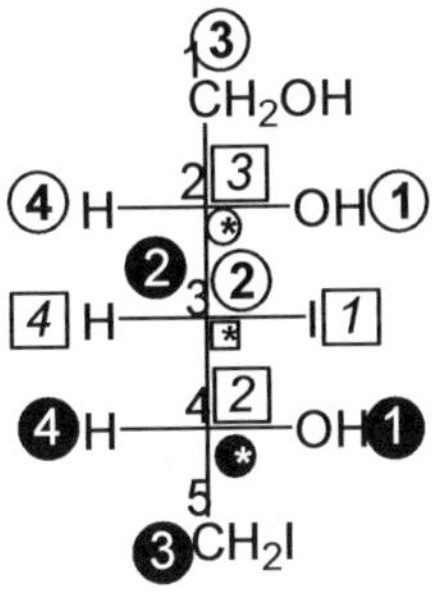

7. Preparation:

   a) ethyl phenyl ether

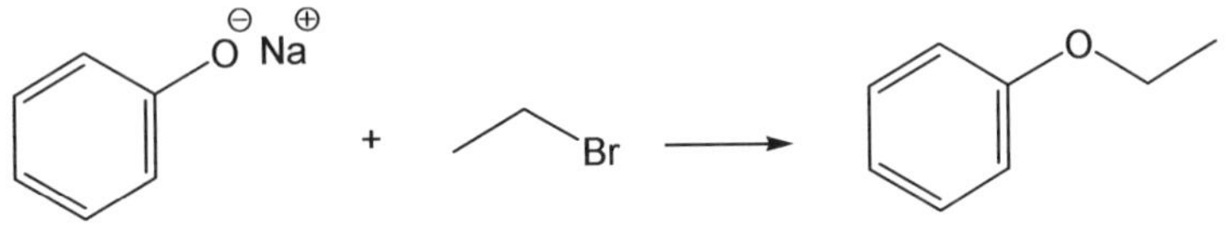

   b) 1-penten-4-yne

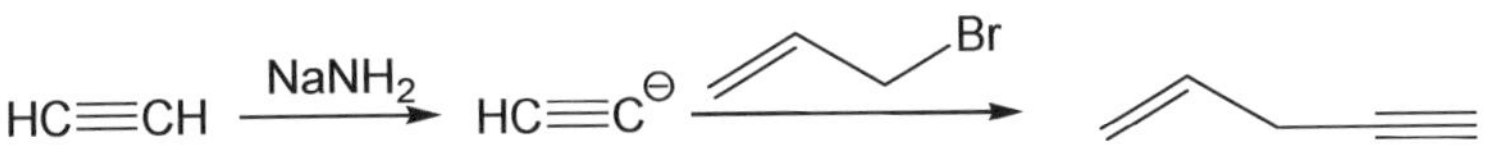

   $HC\equiv CH \xrightarrow{NaNH_2} HC\equiv C^{\ominus} \xrightarrow{\text{allyl bromide}}$ 1-penten-4-yne

8. The formula and name of the product:

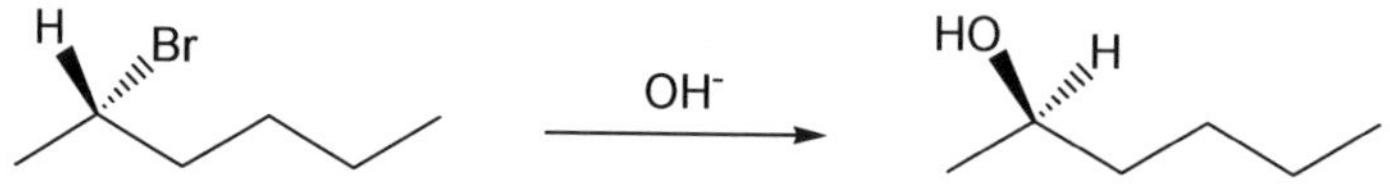

CH₃CH(Br)CH₂CH₂CH₂CH₃ (with H and Br shown on stereocenter) → (with OH⁻) → (R)-2-hexanol (HO and H shown on stereocenter)

9. For each of the following pairs circle the better leaving group:

   a) **(I–)**   or   Cl–

   b) **(H₂O⁺)**   or   HO–

   c) H₂N–   or   **(H₃N⁺)**

10. Mechanism and reaction product:

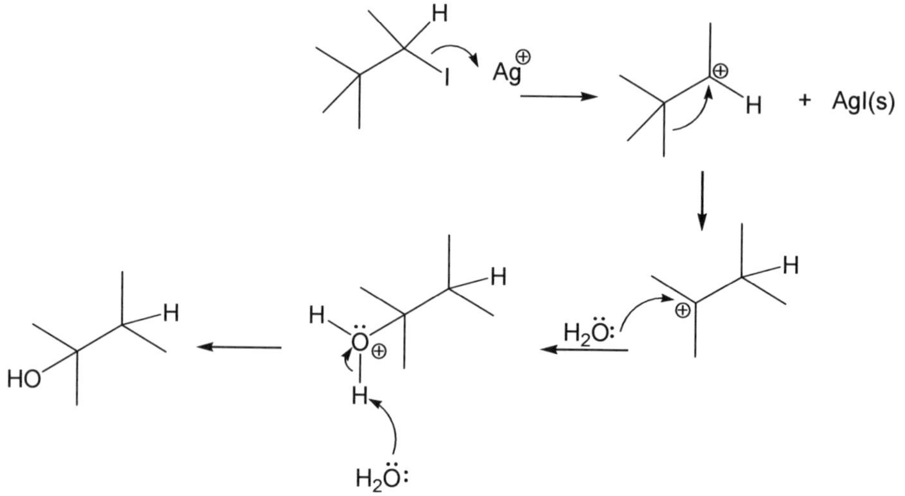

(2,3-dimethyl-2-iodobutane type substrate) + Ag⁺ → carbocation intermediate with hydride shift (H migrates) + AgI(s)

→ rearranged carbocation attacked by H₂O:

→ protonated alcohol intermediate, deprotonated by H₂O:

→ tertiary alcohol product (HO-C(CH₃)₂-CH(CH₃)-H type)

11. Formulas of the missing compounds:

- Cyclohexyl bromide (1-methyl) + H₂O → 1-methylcyclohexan-1-ol
- 3,3-dimethylbutan-2-ol + HBr → 2-bromo-3,3-dimethylbutane (Br on C2)
- trans-1,3-dibromocyclobutane + NaI/acetone → trans-1,3-diiodocyclobutane
- sec-butyl bromide + ⁻SH → sec-butyl thiol (2-butanethiol)
- (E)-(2-bromovinyl)cyclohexane + NaNH₂ → cyclohexylacetylene (C₆H₁₁−C≡CH)

12. A synthesis:

$(CH_3)_3C-O^{\ominus} Na^{\oplus}$ + CH_3CH_2-Br → $(CH_3)_3C-O-CH_2CH_3$

13. Reaction mechanism:

$H_3C-C≡C-CH_3$ →(Na)→ radical anion vinyl intermediate → (H−NH₂) protonation → vinyl radical → (Na) → vinyl anion → (H₂N−H) protonation → (E)-2-butene

Dissolving Metal Reduction

14. A mechanism and the final product:

15. Preparation

16. Complete the following table giving the structure or name for each molecule.

5-methyl-2-hexyne	
3-methyl-1-pentyne	
acetylene	H—C≡C—H

Exercises in Organic Chemistry

13. Elimination Reactions

Concepts to review: E1 and E2 mechanisms, Zaitsev and Hofmann elimination, stereochemistry of an elimination, elimination from cyclic compounds, competition between substitution and elimination.

Questions 1-6.

1. Reaction of 3-iodohexane with KOH can give six products, assuming that there are no rearrangements in the course of the reaction. Provide structural formulas of all six products.

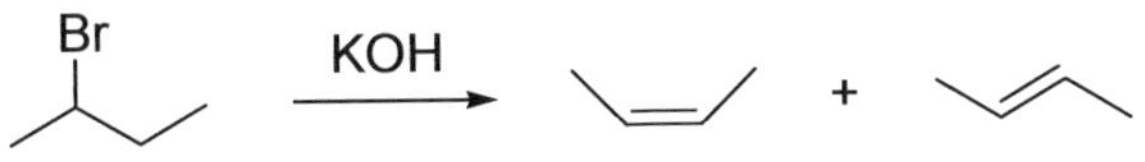

2. In the following example, identify the major product and explain why is it formed in a larger amount?

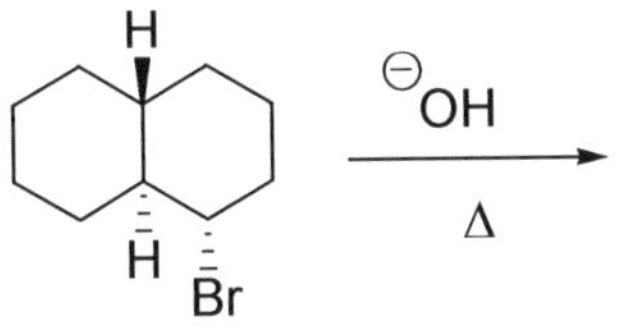

3. What is the mechanism of the following elimination reaction?

   a) E1
   b) E2
   c) May be either E1 or E2
   d) This compound does not undergo elimination under ordinary conditions.

4. Which compound is the major or the predominant product of the elimination reaction shown below?

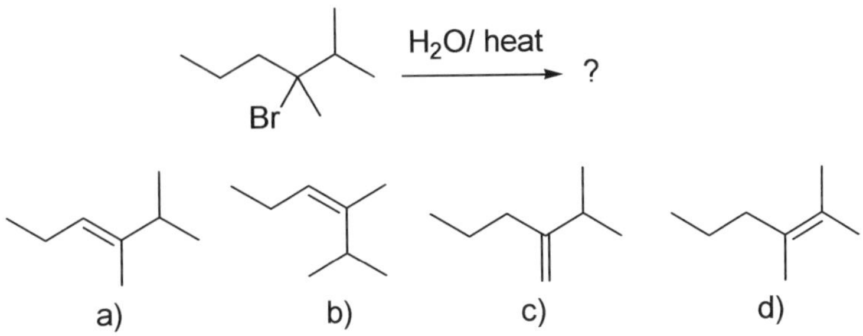

5. Provide formulas of the missing compounds or reagents.

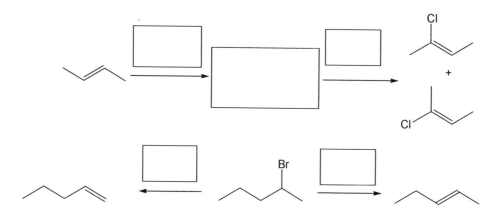

6. Provide the major product for each of the following reactions.

Answers to questions 1-6.

1. Since the mechanism and the reaction conditions have not been specified, one can assume that both substitution as well as elimination reactions are possible. There are two possible substitution products: (*R*)-3-hexanol and (*S*)-3-hexanol. Next, we should examine elimination reaction. There are two possible regioisomers: 2-hexene and 3-hexene. Each of them has two geometric isomers: (*E*)- and (*Z*)-.
Therefore there are a total of two substitution and four elimination products.

2. The reaction mechanism is E2. It is an elimination of a secondary alkyl halide with a strong base – hydroxide ion. In an E2 elimination reaction, stereochemistry of the product is determined by the nature of the transition state. Therefore, the lowest energy transition state will give the major product. In the transition state, alkyl halide has to assume a conformation in which the bonds that are being broken (carbon-halogen and carbon-hydrogen are anti-periplanar to each other). There are two such transition states. Transition state on the left gives *cis*-2-butene as the reaction product, while one at the right gives *trans*-2-butene. Transition state on the left has two methyl substituents next to each other (gauche arrangement) and has a considerable steric strain. In the transition state on the right methyl groups are *anti* to each other and there is no steric strain. Therefore, this is the lower energy transition state and the predominant product is the *trans* isomer.

3. First you should consider what type of bromide are you dealing with? It is a secondary bromide. Therefore, it can eliminate by either E1 or E2 mechanism. Use of a strong base (hydroxide ion) points to E2 mechanism and some students may choose that answer without a further analysis.
    A quick choice of such answer leaves out some important considerations. One has to consider in this particular reaction is that bromine substituent is on a cyclohexane ring. Furthermore, this cyclohexane ring is fused to another in a *trans*-fashion. Therefore, this question is much more complex than it appears at a first glance. It is not a simple question that tests your knowledge on effect of bases on mechanism of elimination reaction, but rather a complex question which requires of you to integrate knowledge of stereochemistry with details of reaction mechanism. This illustrates importance of a correct assessment of a question (whether it is open or closed and a level of difficulty).
    First step in any problem that involves elucidation of a reaction mechanism on a cyclic substrate is to consider the actual conformation of it. Therefore, one should draw the actual conformation or, even better, make a model. Since you may not have a model set during a test, you should learn to

solve problems like these using pen and paper only. The conformation of the molecule is shown on the right.

Now that you know the conformation of the molecule, consider requirements for an E2 reaction. In an E2 reaction carbon-leaving group bond (in this case C-Br bond) and one of the carbon-hydrogen bonds on the neighboring carbon have to be anti-periplanar, or in exceptional cases syn-periplanar.

In this molecule, we can see that only carbon-carbon bonds are anti-periplanar to the carbon-bromine bonds.

Furthermore, you should recall that in a molecule of *trans*-decalin a ring flip to the alternate chair conformation is not possible. If the molecule were a *cis*- stereoisomer, a ring flip would have been possible. Therefore, based on the analysis above one has to conclude that an E2 reaction is not possible. That eliminates two of the possible answers: b) and c). At this point, some students may choose the answer d). Their reasoning goes as follows: "E2 mechanism is not possible due to unfavorable geometry for E2 elimination and since a strong base, which favors E2 elimination, is used, E1 reaction does not happen either. Therefore, the correct answer is d)." A problem with this reasoning is that, while it is true that a strong base favors E2 mechanism, it does not prevent an elimination reaction occurring by E1 mechanism. Therefore, E1 is a possible reaction mechanism. To show if it can be the mechanism of elimination in this case, we should consider the intermediate carbocation. In the intermediate carbocation, there are two carbon-hydrogen bonds that are correctly aligned with the "empty" p orbital for elimination. Therefore, E1 is a plausible reaction mechanism and the correct answer is a).

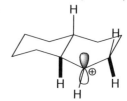

You may note that there is a possibility of a rearrangement into a tertiary carbocation and that there is a possibility for formation of two regioisomers. Whether rearrangement occurs, or whether one of two regioisomers are formed is not important to answer this question. You can try to answer those as an exercise. Keep in mind that the tertiary carbocation that would result from a rearrangement is a not an ordinary tertiary carbocation, but a bridgehead one.

In summary, the correct answer is a) and the distractors are:
b) Designed for students who only analyze problems superficially, have limited knowledge (in this case about conformation of a six memebered ring and how it affects mechanism of elimination), or students who do have such knowledge but are not able to make the necessary connections.
c) Mainly designed for students who did not study. It may tempt some students who have very limited knowledge (for example, those who know only that a secondary substrate may undergo either E1 or E2 reaction, but have no deeper understanding of the mechanisms). Test-smart students usually do not choose an answer like this.
d) This answer is designed for students who have a misconception that since a strong base favors E2 reaction mechanism, an E1 reaction cannot occur in the presence of a strong base. This is another answer that test-smart students usually do not select.

4. A mechanism of this elimination is E1 (weak base and elimination from a tertiary alkyl halide). Therefore, intermediate is the corresponding carbocation and the elimination proceeds to form the most stable, most substituted, alkene. Of the four alkenes, alkenes a) and b) are trisubstituted, alkene c) is disubstituted. The most substituted alkene is d) tetrasubstituted and that is the correct answer.

5. It is the best to work out the first problem by going backwards – from the products to the starting material. The first question is: What starting material and which reagent are needed to produce a *cis/trans* mixture of 2-chloro-2-butene. An elimination of dichloroalkane with a hydroxide gives the desired product. Therefore, reagent is a hydroxide (for example, KOH or NaOH) and the substrate is 2,3-dichlorobutane). Next question is: How to prepare 2,3-dichlorobutane from 2-butene?" The answer to that one should be easy by now. Addition of chlorine will accomplish the desired transformation. Therefore, the reagent is chlorine.

In the second exercise, a single starting material is supposed to give two different products. Obviously, one should use different reagents. Upon closer inspection, it is clear that the product on the left, 1-hexene is a result of Hofmann elimination and the product on the right, 2-hexene, is a result of Zaitsev elimination. Therefore, one should use suitable bases. A strong base, with a small anion such as hydroxide, favors Zaitsev elimination, while a large hindered base, such as *t*-butoxide, favors Hofmann elimination. Therefore, those are the two reagents.

6. Reaction products:

The only reasonable elimination product. Allene is too high in energy and is not a reaction product.

Formation of a conjugated diene as a more stable molecule than an isolated diene.

Bulky base - Hofmann elimination.

Zaitsev elimination

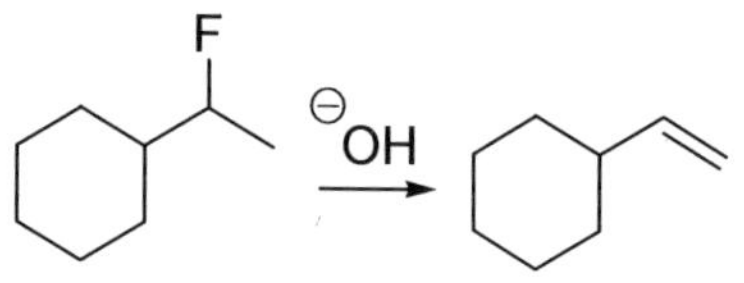

[cyclohexyl-CHF-CH3] →(⁻OH) [cyclohexyl-CH=CH2]   Poor leaving group - Hofmann elimination.

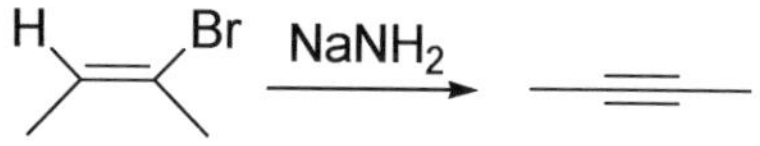

(CH3)(H)C=C(Br)(CH3) →(NaNH2) CH3-C≡C-H

Additional Exercises: Questions 7-16

7. Predict the major product of each of the following reactions.

   CH3CH2-CH(Br)-CH2-CH=CH2 →(CH3O⁻)

   1-bromo-1-cyclohexyl →(H2O, Δ)

   (CH3)2CH-CH(Br)-CH3 →(KOH, heat)

   (CH3)2CH-CH(F)-CH3 →(CH3O⁻, E2)

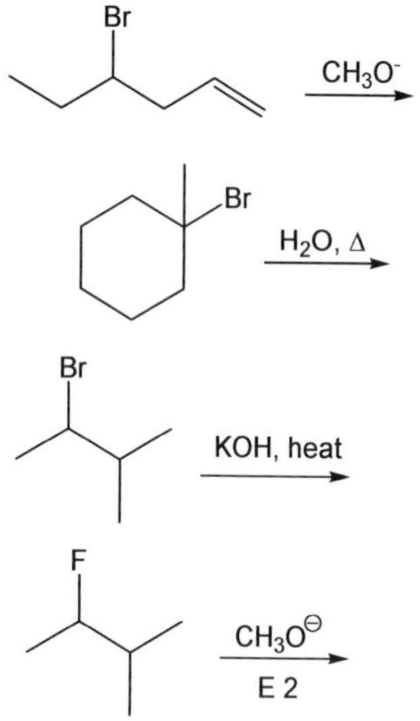

8. Provide a starting material for each of the following reactions.

   ? →(NaNH2, 1 equivalent) cyclohexyl-C≡CH

   ? →(KOH) CH3CH2CH2-C(Br)=CH2 [product drawn as CH2=C(Br)-CH2CH2CH3]

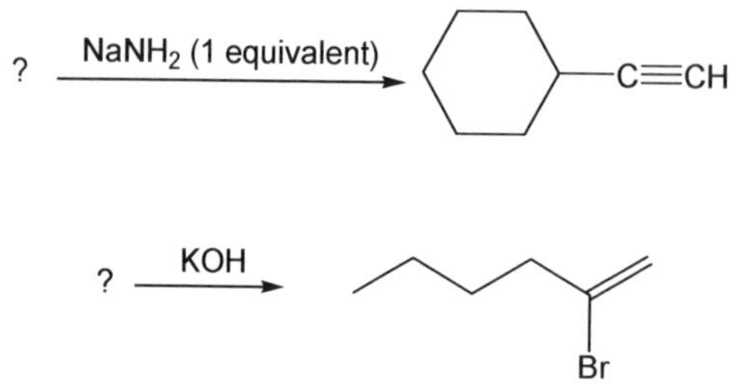

9. Provide the reagent for each of the following reactions.

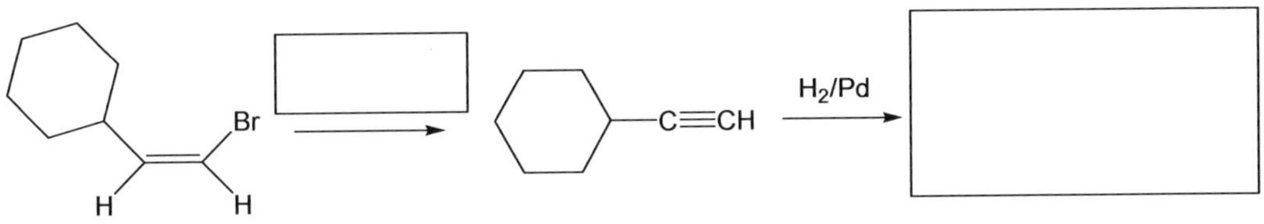

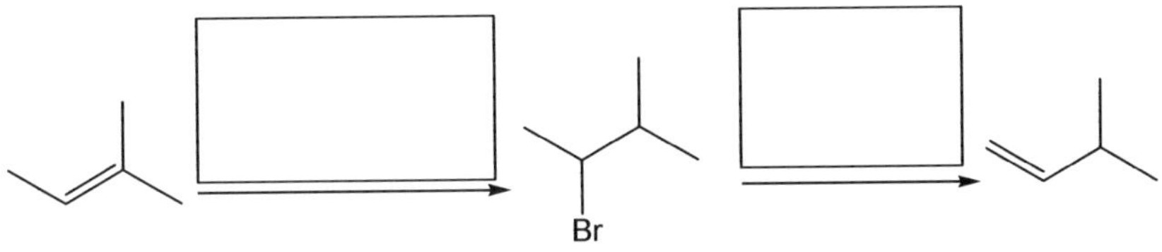

10. Elimination reaction of 3,3-dimethyl-2-bromobutane in boiling methanol gives 2,3-dimethyl-2-butene. Which of the following is the best description of this reaction?

    a) Concerted reaction which gives the most stable alkene.
    b) Concerted reaction which proceeds via the lowest energy transition state.
    c) Carbocation formation, a 1,2-alkyl shift and loss of a proton.
    d) Carbocation formation, a 1,2-hydrogen shift and loss of a proton.

11. How many products can be formed in an elimination reaction of 2-chloropentane?

    a) 2
    b) 3
    c) 4
    d) 6

12. Vinyl bromides undergo elimination reaction when treated with:

    a) NaNH$_2$
    b) H$_2$SO$_4$
    c) Na in liquid NH$_3$
    d) HBr or HI

13. What is the major product of the following reaction and what is the reaction mechanism?

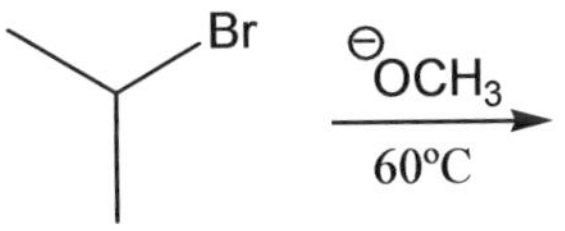

    a) isopropyl methyl ether; S$_N$1
    b) isopropyl methyl ether; S$_N$2
    c) propene; E1
    d) propene; E2

14. In a reaction of a 2-bromo-2-pentene and sodium amide the product is:

    a) 2-pentyne
    b) 2-pentene
    c) 1-pentyne
    d) 1-pentene

Questions 15 and 16: Menthyl chloride, shown below, undergoes elimination upon treatment with potassium hydroxide to give the alkene below as the only reaction product.

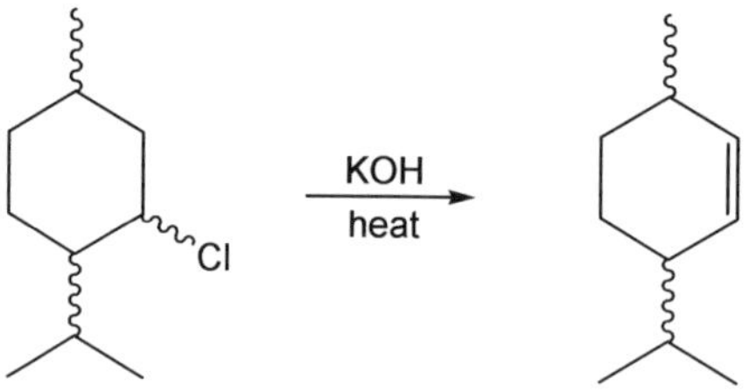

15. What is the reaction mechanism?

    a) E1
    b) E2
    c) Either E1 or E2 is a possible reaction mechanism since both give the same reaction product.
    d) Free radical elimination

16. Chloro- and isopropyl substituents are:

    a) *cis* to each other
    b) *trans* to each other
    c) *syn* to each other
    d) eclipsing each other

Answers to Additional Exercises: Questions 7-16

7. Reaction products:

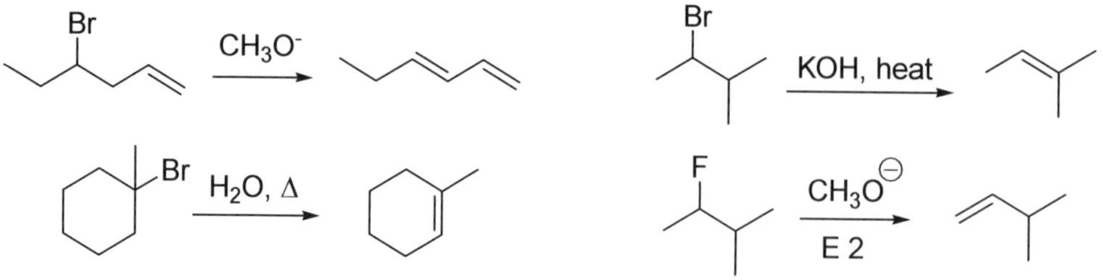

8. Starting materials:

[Reaction 1: cyclohexyl-CH=CHBr (either cis or trans isomer) + NaNH₂ (1 equivalent) → cyclohexyl-C≡CH]

[Reaction 2: 2,3-dibromohexane + KOH → 2-bromo-1-hexene]

9. Missing compounds:

[Reaction 1: cyclohexyl-CH=CHBr + NaNH₂ → cyclohexyl-C≡CH + H₂/Pd → cyclohexyl-CH₂CH₃ (ethylcyclohexane)]

[Reaction 2: 2-methyl-2-butene + HBr, peroxides → 2-methyl-3-bromobutane; then + KOC(CH₃)₃ → 3-methyl-1-butene]

10. c); 11. b); 12. a); 13. d); 14. a); 15. b); 16. b).

Veljko Dragojlovic

14. Substitution and Elimination Reactions of Alcohols and Ethers

Concepts to review: S_N, S_N2, E1 and E2 reactions of alcohols, Lucas test, phosphorus halides and thonyl chloride, sulfonate esters, reactivity of epoxides (oxiranes), , organometallic reagents.

Questions 1-8.

1. Provide formulas of all the possible products in a reaction of 3-methyl-2-butanol with HCl?

2. How many isomers could theoretically be obtained by an acid-catalyzed elimination of water from 2-butanol?

    a) None. Alcohols do not eliminate water under acidic conditions.
    b) 2
    c) 3
    d) 4

3. Why HBr and HI cleave ethers, while HCl does not?

4. Compare reactions of phosphorus trichloride and phosphorus oxychloride with an alcohol?

5. Why are amines poor leaving groups?

6. What is the polarity of carbon atom in an alkyl halide and the corresponding organometallic compound?

7. What is the structure of a Grignard reagent? Why it has to be prepared in ether as a solvent?

8. What is the oxidation state of copper in a Gilman reagent?

Answers to questions 1-8.

1. An alcohol can react with HCl in either a substitution or an elimination reaction. There are a number of possible substitution products: (*R*)-3-methyl-2-butanol and (*S*)-3-methyl-2-butanol. The major substitution product is most likely going to be a rearrangement product 2-methyl-2-butanol. Next, we should examine elimination reaction. There are two possible regioisomers: 3-methyl-1-butene and 2-methyl-2-butene. Therefore there are a total of three substitution and two elimination products.

2. This is an open-ended question and the best way to answer it is to do so without looking at the offered options and then match your answer with one of the options. The possible elimination products are 1-butene, *cis*-2-butene and *trans*-2-butene and the correct answer is c).

3. To cleave an ether, it is not enough to convert oxygen into a good leaving group by complexing it with an electrophile (proton). In addition, a strong nucleophile is required. Role of the electrophile is to "pull" oxygen (the leaving group, nucleophile) out of the molecule, while role of the strong nucleophile is to "push" it out. For a reaction on an unreactive molecule such as ether to occur, both a "pull" by electrophile and a "push" by nucleophile are needed. This is called a "push-pull effect." All three acids are strong acids and are capable of protonating ethers. However, only bromide and iodide ions are strong enough to displace an oxonium ion. Chloride ion is much weaker nucleophile and cannot displace an oxonium ion.

4. Both reagents react with an alcohol to give a similar intermediate – oxonium ion. An oxonium ion is a good leaving group and in a reaction with a nucleophile it undergoes a nucleophilic substitution reaction (S_N2), while in a reaction with a base it undergoes E2 elimination reaction. Reaction between an alcohol and phosphorus trichloride generates chloride ions, which act as nucleophiles and the overall reaction is a substitution reaction. In a reaction between an alcohol and phosphorus oxychloride, pyridine is an

additive and acts as a base. It deprotonates the intermediate oxonium ion, which makes oxygen substituent into a weaker leaving group. Another molecule of pyridine acts as a base (pyridine is non-nucleophilic) and attacks a weakly acidic hydrogen on the neighboring carbon. The overall reaction is E2 elimination.

5. In an amine, nitrogen forms a strong bond with carbon. Such bond is difficult to break. Once it is broken the resulting leaving group (amide anion) is a strong base. As a strong base it has a high tendency to re-form bond to the carbon. That is why strong bases make poor leaving groups. Nitrogen atom can be made into a somewhat better leaving group by converting it into an ammonium ion (for example, by protonation). However, even ammonium ion, upon breaking of carbon-nitrogen bond, gives a base - the corresponding amine. Thus, although an ammonium ion is a better leaving group than nitrogen itself, it is not a very good leaving group either.

6. In an alkyl halide carbon atom has a partial positive charge, while in an organometallic reagent it has a partial negative charge. Thus, in the course of preparation of an organometallic reagent from an alkyl halide, polarity of a carbon atom changes from positive to negative and from an electrophile carbon atom becomes a nucleophile.

$$\overset{\delta+}{H_3C}—\overset{\delta-}{Cl}$$
carbon is an electrophile

$$\overset{\delta-}{H_3C}—\overset{\delta+}{Li}$$
carbon is a nucleophile

7. In a Grignard reagent carbon–magnesium bond is a polar covalent bond with a partial negative charge on the carbon atom, while magnesium-bromine bond is an ionic bond. A Grignard reagent reacts as a carbon nucleophile. Therefore, a carbon atom reacts as if the carbon–magnesium bond were ionic and the carbon atom were a carbanion. Thus, we often represent the structure of a Grignard reagent as shown below, with an electronic pair and negative charge on the carbon atom and positive charge on the MgBr residue (we are not concerned with detailed structure of that residue), even though we know that it is not a correct representation – such representation helps us predict the outcome of reactions of Grignard reagents.

$$\overset{\delta-}{CH_3CH_2}—\overset{\delta+}{Mg}\overset{\oplus}{}\overset{\ominus}{Br}$$
actual structure

$$CH_3\overset{\ominus}{\underset{..}{C}H_2} \overset{\oplus}{MgBr}$$
usual representation

A Grignard reagent has to be prepared in ether because it is soluble in that solvent. It is insoluble in common non-polar organic solvents such as hexane or toluene.

8. Gilman reagent is a lithium dialkylcuprate. Lithium is a cation and dialkylcuprate is an anion. Since lithium ion always has a formal charge of +1 (it is group I metal), dialkylcuprate anion has a charge of -1. The charge on copper depends on the nature of carbon-copper bond. That bond is polarized the same way as in other organometallic reagents – there is a partial negative charge on the carbon and a partial positive charge on copper. Therefore, oxidation states are -1 for the carbon atom and +1 for the copper atom.

R_2CuLi
lithium diethylcuprate

$(CH_3)_2CuLi$
lithium diethylcuprate

$$\overset{\delta-}{H_3C}—\overset{\delta+}{Cu}—\overset{\delta-}{CH_3}$$

Questions 9 and 10.

9. Provide the only or the major products of the following reactions.

cyclohexyl-OH + Cl-S(=O)-Cl / pyridine → ☐

(CH₃)₃C-CH₂-OCH₃ (neopentyl methyl ether structure shown) + HBr →Δ→ ☐

cyclopentyl-OH + PI₃ → ☐

CH₃CH₂Li + (epoxide, ethylene oxide) → ☐ →H₃O⁺→ ☐

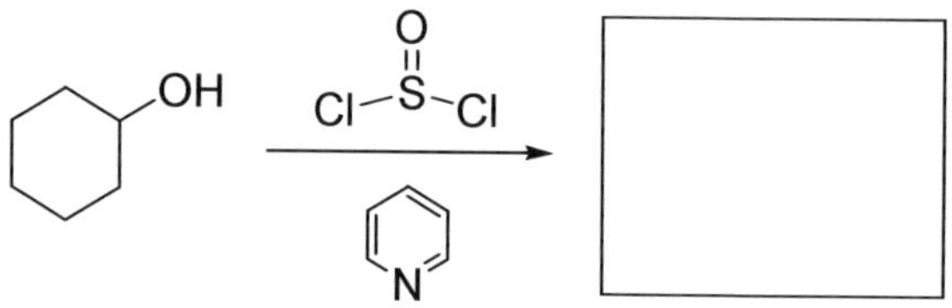

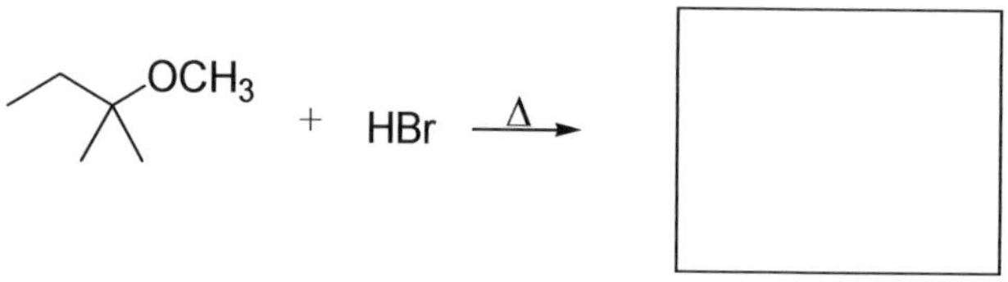

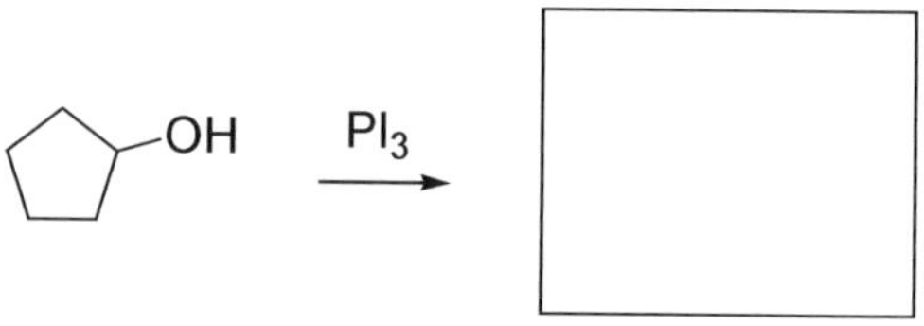

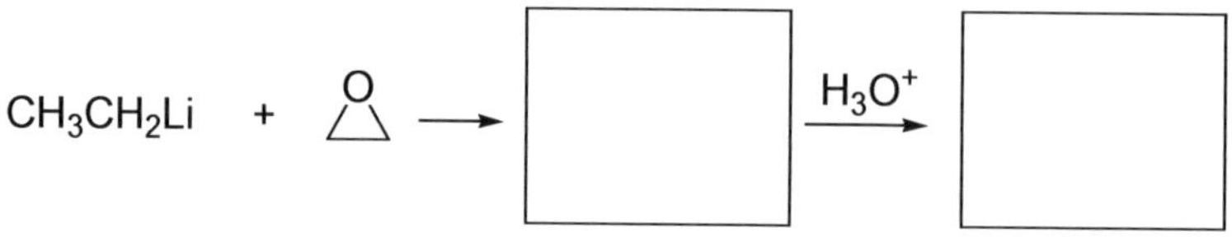

10. Provide formulas of the reagents for the following reactions.

(cyclohex-3-enyl)-CH₂-OH → ☐ → (cyclohex-3-enyl)-CHO

cyclohexyl-OH → ☐ → cyclohexyl-I

cyclohexene oxide → ☐ → trans-2-methylcyclohexan-1-ol (OH and CH₃ trans)

cyclopentyl-S-CH₃ → ☐ → cyclopentyl-S⁺(CH₃)₂ Br⁻

142

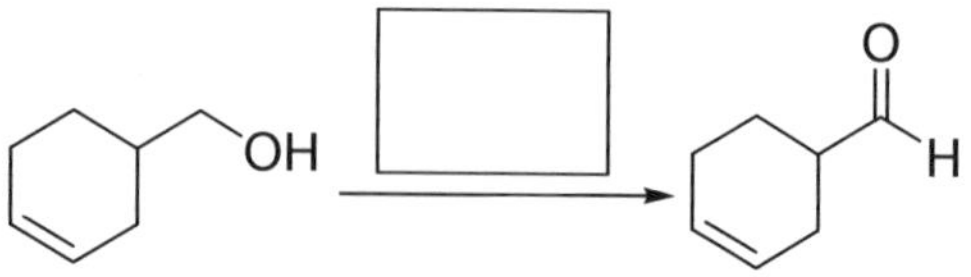

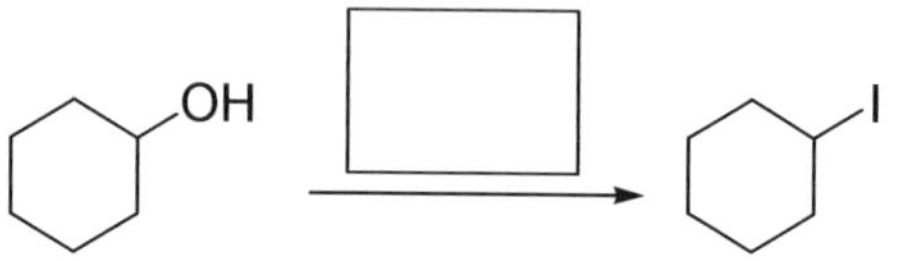

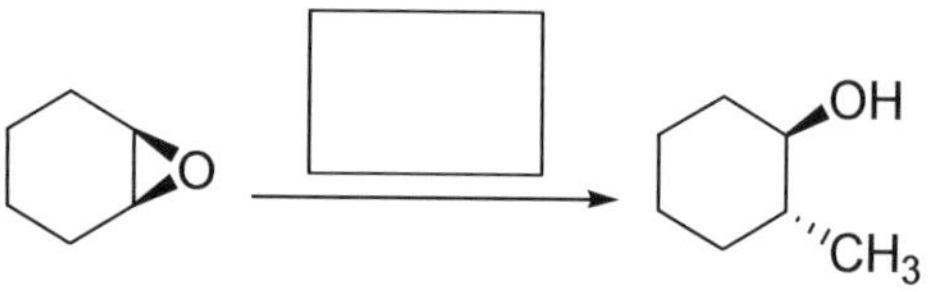

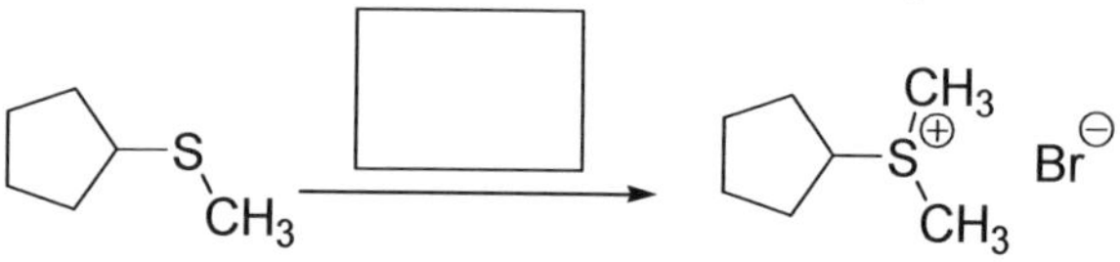

Answers to questions 9 and 10.

9. Reaction products:

Cyclohexanol + Cl-S(=O)-Cl / pyridine → cyclohexyl chloride

2-methoxy-2-methylbutane + HBr, Δ → 2-bromo-2-methylbutane + CH$_3$OH

Cyclopentanol + PI$_3$ → cyclopentyl iodide

CH$_3$CH$_2$Li + ethylene oxide → CH$_3$CH$_2$CH$_2$CH$_2$O$^-$Li$^+$ →(H$_3$O$^+$)→ 1-butanol

10. Reagents:

(cyclohex-3-enyl)methanol →[PCC, CH$_2$Cl$_2$]→ (cyclohex-3-enyl)carbaldehyde

Cyclohexanol →[P$_4$, I$_2$ or PI$_3$]→ cyclohexyl iodide

Cyclohexene oxide →[1) CH$_3$Li; 2) H$_3$O$^+$, H$_2$O]→ trans-2-methylcyclohexanol

Cyclopentyl methyl sulfide →[CH$_3$Br]→ cyclopentyl dimethyl sulfonium bromide

143

Additional Exercises: Questions 11-20

11. Provide formulas of the missing compounds.

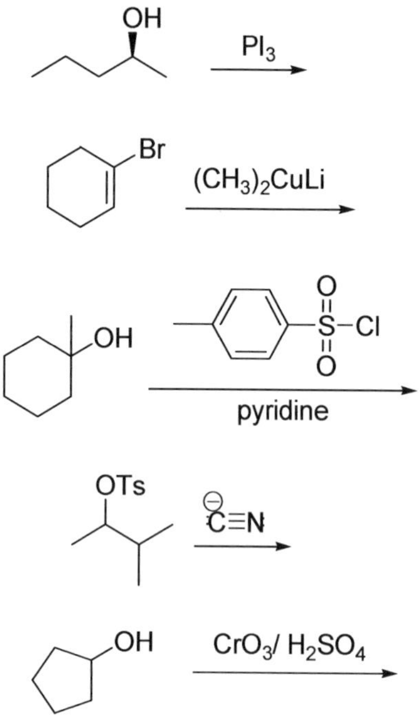

[Structure: 2-butanol (CH₃CH₂CH(OH)CH₃)] $\xrightarrow{PI_3}$

[Structure: 1-bromocyclohexene] $\xrightarrow{(CH_3)_2CuLi}$

[Structure: 1-methylcyclohexanol] + [tosyl chloride structure] $\xrightarrow{\text{pyridine}}$

[Structure: 3-methyl-2-butyl tosylate] $\xrightarrow{{}^{\ominus}C\equiv N}$

[Structure: cyclopentanol] $\xrightarrow{CrO_3/\ H_2SO_4}$

12. Provide reagents for the following reactions. Note that some transformations may involve several steps.

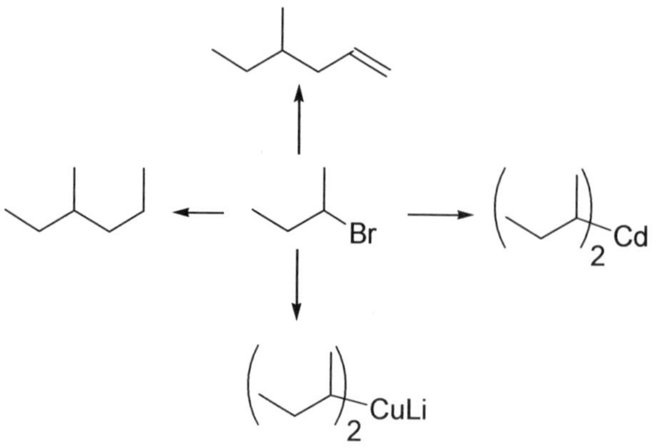

[Scheme showing a central secondary alkyl bromide transforming into: 4-methyl-1-pentene (up), 2-methylpentane (left), (R)₂Cd (right), and (R)₂CuLi (down)]

13. The first step in a substitution reaction of an alcohol is:

  a) hydration
  b) formation of an oxonium ion
  c) formation of a hydronium ion
  d) deprotonation

14. Primary alcohols do not undergo S_N1 reaction because:

    a) primary carbocations are unstable
    b) primary carbocations are unreactive
    c) the enthalpy of formation is positive
    d) the entropy of formation is positive

15. The following elimination reaction might produce the following products. Which compound is the major or the predominant product?

    [Structure: alcohol with OH, H₂SO₄ (conc.) → ?]

    a)   b)   c)   d)

16. Alcohols are readily dehydrated with:

    a) reducing agents
    b) oxidizing agents
    c) acids
    d) bases

17. Alcohols undergo elimination reaction when treated with:

    a) PCl_3
    b) Na in liquid NH_3
    c) $POCl_3$ and pyridine
    d) $SOCl_2$ and pyridine

18. Which of the following is a good leaving group?

    a) $-OH_2^+$
    b) $-OCH_3$
    c) $-N(CH_3)_3^+$
    d) $-SCH_3$

19. Alkyllithium compounds undergo transmetallation reaction with:

    a) H_2O
    b) ether
    c) CuI
    d) $SOCl_2$ and pyridine

20. Ethers undergo substitution reaction when treated with:

    a) PCl_3
    b) $SOCl_2$
    c) $POCl_3$ and pyridine
    d) HI

Answers to Additional Exercises: Questions 11-20

11. Reaction products:

(2-pentanol) + PI₃ → 2-iodopentane (with inverted stereochemistry)

1-bromocyclohexene + (CH₃)₂CuLi → 1-methylcyclohexene

1-methylcyclohexanol + TsCl / pyridine → 1-methylcyclohexyl tosylate

(3-methylbutan-2-yl) OTs + ⁻C≡N: → 3-methyl-2-cyanobutane (nitrile)

cyclopentanol + CrO₃ / H₂SO₄ → cyclopentanone

12. Reagents:

- (CH₃CH₂CH₂)₂CuLi, ether → 2-methylhexane (from 2-bromobutane starting material shown centrally)
- (CH₂=CHCH₂)₂CuLi, ether → 3-methyl-1-hexene (allyl coupling product)
- 1) Mg, ether; 2) CdCl₂, ether → R₂Cd
- 1) Li, hexane; 2) CuI, ether → R₂CuLi

13. b); 14. a); 15. b); 16. c); 17. c); 18. a); 19. c); 20. d).

Exercises in Organic Chemistry

16. Spectroscopy

Concepts to review: Identification of Organic Compounds, Mass Spectrometry (applications of MS, HRMS and LRMS, Fragmentation Pattern), Infrared (IR) Spectroscopy (absorptions of common functional groups), Ultraviolet (UV) Spectroscopy MO Interpretation of UV spectroscopy, Nuclear Magnetic Resonance (NMR) (Shielding and The Chemical Shift, The Number of Signals in the ^1H NMR Spectrum; Spin-Coupling/Splitting, Nonequivalent Nuclei, Integration of the NMR Signals, Interpretation of ^1H NMR Spectra, ^{13}C NMR Spectroscopy, Interpretation of ^{13}C NMR Spectra, Symmetry and NMR spectra).

Questions 1-5.

1. Which of the following compounds has m/z of 84?   *mass spec*

   a) cyclopentyl-OH   b) cyclopentanone (=O)   c) cyclopentyl-SH   d) cyclopentyl=S

2. How many signals are in ^{13}C NMR spectrum of the molecule shown below?

   $H_3C\underset{e}{}\underset{d}{CH_2}\underset{c}{CH_2}\underset{b}{CH}(OCH_3)(OCH_3)$ — labels: e, d, c, b, a ; OCH$_3$ labels a

   a) 3
   b) 4
   c) 5   ⟵ (circled)
   d) 6

3. A proton NMR spectrum of ethyl methyl ether shows:

   a) three peaks, a singlet, a triplet, and a doublet.
   b) three peaks, a singlet, a triplet, and a quartet.   ⟵
   c) three peaks, two triplets, and a doublet.
   d) three peaks, two triplets, and a quartet.

   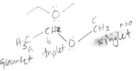
   H$_5$ ^{a} — CH$_2$ ^{b} — O — CH$_3$ ^{c}
   quartet   triplet   singlet

4. Predict the multiplicity in a proton NMR spectrum for each underlined proton in the following compounds. ^1H-NMR   [multiplicity = n+1]   [n = neighboring protons]

   H—C(H)(H)—C(H)(H)—C(=O)—H   CH$_3$CH$_2$CHO   (quartet of triplets / triplet of quartets)   CH$_3$CH$_2$CH$_2$OCH$_3$ n=2 ⟹ triplet   Ph—CH$_2$—CH$_3$ quartet
   n=2 triplet   n=2 triplet   n=3 ?

5. Predict the multiplicity (the number of peaks as a result of splitting) for each underlined carbon in the following compounds.   [n = protons on C atom]   [n+1 = multiplicity]

   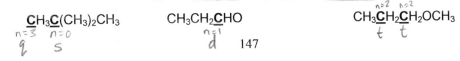
   CH$_3$C(CH$_3$)$_2$CH$_3$ CH$_3$CH$_2$CHO   CH$_3$CH$_2$CH$_2$OCH$_3$
   n=3 n=0   n=1   n=2 n=2
   q   s   d   t   t

147

Answers to questions 1-5.

1. To answer this question, one has to know what m/z is. Abbreviation m/z stands for "mass (m) to charge (z)" ratio. Therefore, question is related to mass spectrometry. Charge of an ion in a mass spectrometer is +1. Therefore, the mass to charge ratio is actually a mass of the compound and 84 is molar mass of the compound. Of the four possible answers, compound b) cyclopentanone has mass of 84 and is the correct answer.

2. Each signal in ^{13}C NMR spectrum corresponds to a different type of carbon atoms. Carbon atoms in identical magnetic environment give only one signal. Different carbon atoms have different substituents and bonding pattern. A best way to identify all different carbon atoms and count them is to assign a letter to each type of carbon atom. In the molecule of 1,1-dimethoxybutane there are five different carbons (labeled as a-e) and, therefore there are five signals in ^{13}C NMR spectrum or c).

$$H_3C\overset{e}{}\underset{CH_2}{\overset{d}{}}\overset{c}{CH_2}\underset{CH}{\overset{b}{}}\overset{a}{OCH_3}$$
$$\underset{OCH_3}{\overset{a}{|}}$$

3. Ethyl methyl ether has a formula: $CH_3CH_2OCH_3$. Therefore, there are three types of protons and three signals. Each signal is split into a multiplet by protons on the neighboring carbon atom according to a formula: ***multiplicity = n + 1***, where ***n*** is the number of protons on the neighboring carbon atoms. Methyl group bonded directly to oxygen atom has no neighboring protons and ***n*** = 0. Therefore, its signal is a singlet. Ethyl group is the other substituent. Its methyl group is bonded to a methylene group (-CH_2-) and is split into a triplet (***n*** = 2; 2+1 = 3). Finally, methylene group is next to a methyl group and is split into a quartet (***n*** = 3; 3+1 = 4). Therefore, signals are a singlet, a triplet and a quartet, or b).

4. One should follow the procedure described for the question 3. In case of propanal, CH_3CH_2CHO, each of the underlined protons is next to a methylene group and each will be a triplet (abbreviated as *t*). In case of 1-methoxypropanal, $CH_3CH_2CH_2OCH_3$, protons on the carbon 1 are next to a methylene group and the signal is a triplet (*t*). Protons on the carbon 2 are next to a methylene and a methyl group. Each of those two neighboring substituents splits the signal independently and the pattern is either a triplet of quartets (*tq*) or a quartet of triplets (the actual pattern depends on the magnitudes of coupling constants). Most NMR spectrometers cannot resolve such complex signals and we usually refer to such signal as a multiplet (*m*). In ethylbenzene, methylene group is next to a methyl group and a benzene carbon, which has no protons. Therefore, the signal is a quartet (*q*).

$$\overset{t}{H_3C}-\overset{t}{CH_2}-CHO \qquad \overset{tq\text{ or }m}{H_3C}-\overset{t}{CH_2}-CH_2-OCH_3 \qquad \underset{}{\overset{q}{CH_2}}-CH_3$$

5. This question is similar to the previous one. The difference is that carbon atoms are coupled to, and their signal is split by, the protons that are bonded to that carbon atom – not the protons on the neighboring carbons! Therefore, one counts protons on each carbon and calculates multiplicity for the formula n+1:

$$\overset{q}{C}H_3\overset{s}{C}(CH_3)_2CH_3 \qquad CH_3CH_2\overset{d}{C}HO \qquad CH_3\overset{t}{C}H_2\overset{t}{C}H_2OCH_3$$

Questions 6-9.

6. A proton NMR spectrum of compound with a molecular formula C$_3$H$_6$O is given below. Which of the structures shown on the right is the NMR spectrum consistent with? H+

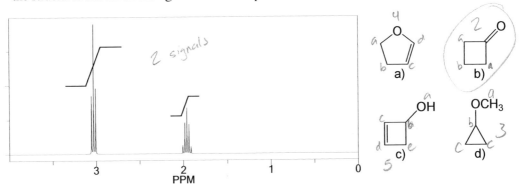

2 signals

7. Which spectroscopic technique would you use to distinguish between these two compounds?

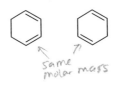

same molar mass

a) IR
b) MS (mass spect)
c) UV-Vis
d) NMR

8. What is the product of the following reaction?

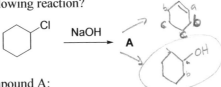

Proton NMR Spectrum of the compound A:

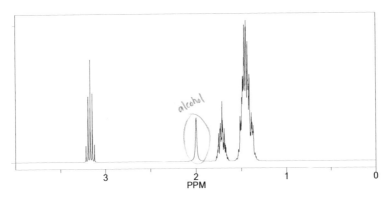

alcohol

149

9. Which of the following compounds exhibit absorption in the UV-Vis part of the spectrum?

a)   b)   c)   d)   e)

Answers to questions 6-9.

6. A question like this has to be treated as a closed question. A single spectrum (NMR. IR or MS) usually does not provide enough information to assign a structural formula to a compound. Therefore, one should eliminate the wrong answers until there is one correct answer left. The key information here is the number of signals in the NMR spectrum. There are only two – one at ~1.95 and the other one at ~3.05. If we examine the four possible choices and the number of expected signals:

We can see that only the compound b) is expected to exhibit two signals in the proton NMR spectrum and that is the correct answer. Multiplicity (a triplet for protons *a*, and a quinted for the protons *b*) confirms that this is the correct answer.

7. In hands of an experienced analytical chemist any of the four techniques would be useful. However, some are better than the others and handling multiple choice questions is all about choosing the best answer. One that can be eliminated easily is MS, since both compound have the same molar mass and the fragmentation patterns of the two are not very different. Of the remaining three, the best one is UV-Vis because 1,3-cyclohexadiene exhibits an absorption and that part of the spectrum, while 1,4-cyclohexadiene does not. Thus, while IR and NMR may also be useful, UV-Vis is the best as it provides a cleat difference between the two and the correct answer is c).

8. To answer this question one has to consider 1) What are the possible reaction products and 2) What are the expected spectra of each of the products. As far as reaction products are concerned, a reaction between an alkyl halide and a hydroxide may give either a substitution or an elimination product, depending on the reaction conditions. Thus, the two possible products are:

The NMR spectrum is consistent with an alcohol and that is the correct answer. There are no signals due to alkene protons (one would expect them in the range 5-6.5 ppm) and there are signals expected for –OH (~2 ppm) and C-H next to –OH (~3.2 ppm).

9. Energy of the ultraviolet and visible part of the electromagnetic spectrum is sufficient to promote electrons from a π into a π^* orbital, or from a n (a non-bonding orbital, which is associated with free electrons) into π^* orbital. Only conjugated multiple bonds have π, π^* sufficiently small energy gap to absorb in UV-Vis part of the spectrum. Isolated double bonds have π, π^* energy gap that is too large and absorbs in so-called far-UV, which is not detected by common UV-Vis spectrometers. Therefore, only compounds with conjugated multiple bonds and compounds with free electron pairs on one atom of a multiple bond absorb in UV-VIs. In the list above those compounds are c) conjugated double bonds and d) free electron pair on oxygen, which is also a part of a double bond.

Questions 10-13.

10. The following reaction was monitored by ^{1}H NMR spectroscopy:

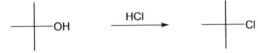

Explain the NMR spectra shown below.

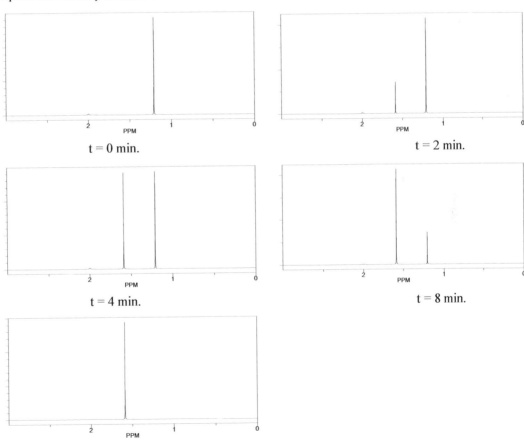

11. Match each IR spectrum shown on the next page with the formula of the compound.

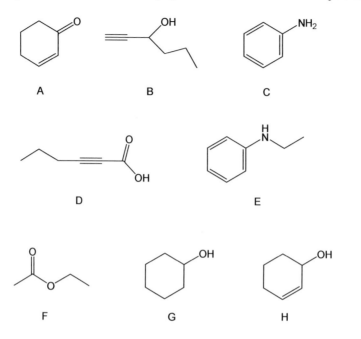

Exercises in Organic Chemistry

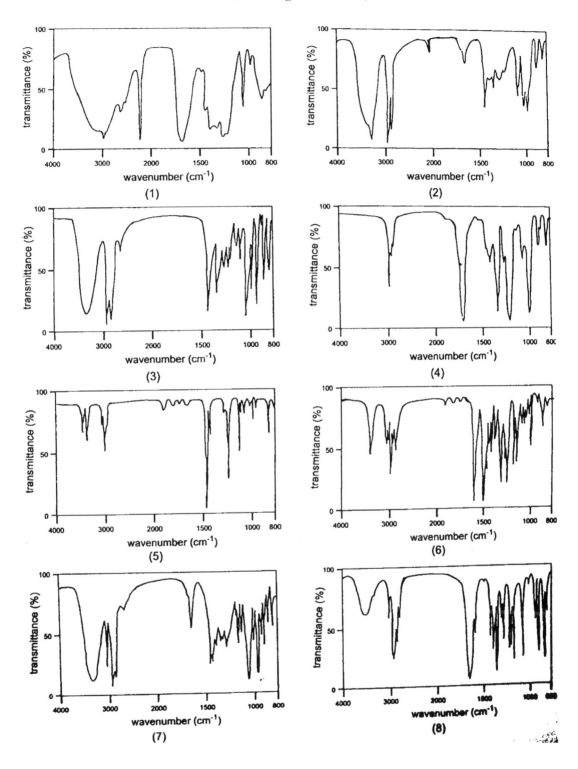

12. Provide a structural formula consistent with the following spectra.

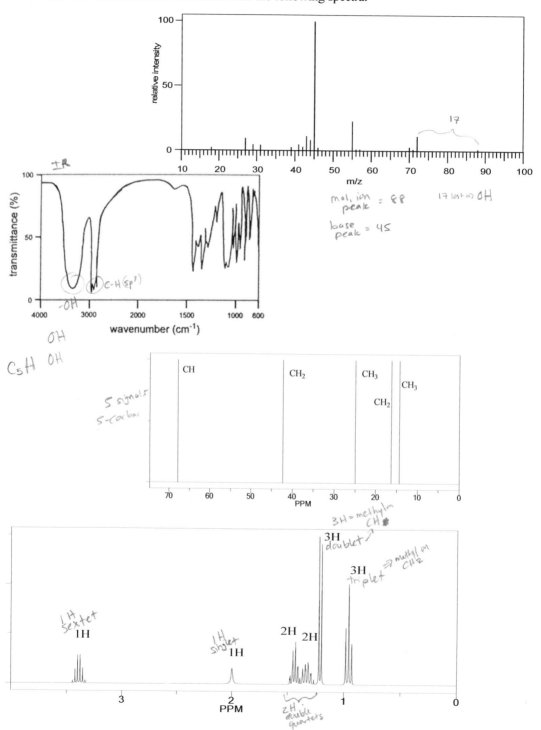

13. Provide a structural formula consistent with the following spectra.

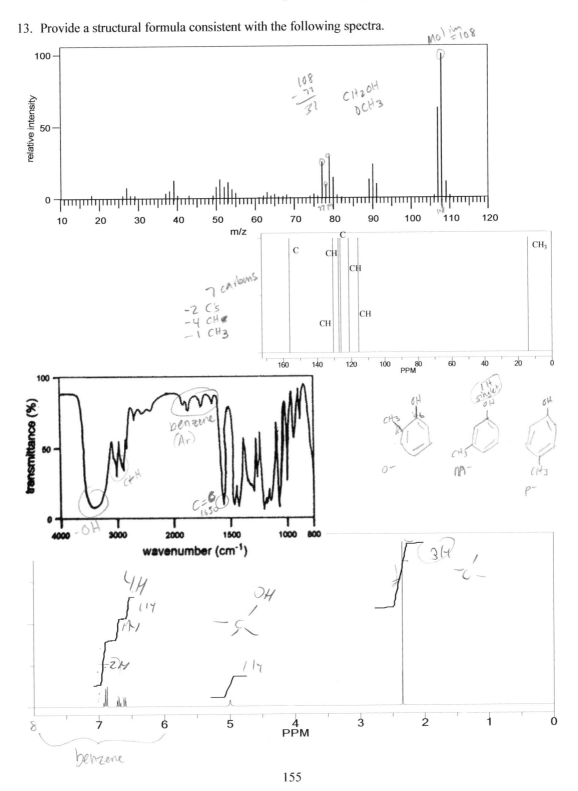

Answers to questions 10-13.

10. At t = 0, the reaction did not start yet and we have only the starting material. This is a spectrum of *t*-butanol. Note the small peak due to –OH proton ~2 ppm. As the reaction progresses, the signals due to *t*-butanol gradually diminish in intensity and signals due to *t*-butyl chloride appear and increase in intensity. Finally, the fact that at t = 16 min, there is only a signal due to *t*-butyl chloride and that there is no signal due to *t*-butanol indicates that the reaction has been completed.

10. The most important mistake that students make when approaching a problem like this is to attempt to solve it chronologically. In other words, they either start with the formula A, try to assign it a spectrum, then the formula B and so on, or they start with the spectrum (1) try to assign it a formula, next the spectrum (2) and so on. The best approach is to assign first the easy ones and then progressively more and more difficult ones. One good starting point is cyclohexanol G. It has only – OH functional group which is expected to exhibit a broad strong absorption ~3300 cm^{-1}. Among the spectra, (1), (2), (3), (7) and (8) have absorption in that region. However, spectra (1), (7) and (8) also have absorptions in the 2000 – 1600 cm^{-1} region, which is where double bonds absorb. Therefore, these can be eliminated. Spectrum (2) has an absorption ~2100 cm^{-1}, which is indicative of an alkyne and can be eliminated, too. This leaves as with the spectrum (3) for cyclohexanol G. We just identified the spectrum (2) as that of a compound having –OH and –C≡C– groups. That is consistent with the formula (B). Another relatively simple compound it the ester (F). One would expect to see only the carbonyl absorption (~1700 cm^{-1}) and possibly C-O bond (~1100 cm^{-1}), which is not considered to be a good diagnostic signal. The spectra with strong sharp carbonyl signals are (1), (4) and (8). Spectrum (1) also has and –OH absorption as mentioned earlier and spectrum (8) has a similar absorption in that area. Therefore, spectrum (4) is that of the compound F. Based on our analysis so far, it is cleat that the spectrum (1) has a carbonyl (C=O) group and an –OH group. This is consistent with the formula D, which also has a triple bond as shown by a strong signal at ~2100 cm^{-1}. Note also that –OH absorption has been shifted to ~3000 cm^{-1} and is overlapping with C-H absorptions, which is also characteristic of a carboxylic acid. Among the remaining spectra, (5) and (6) are very similar in appearance. They have several weak absorptions in the region of ~2000 – 1600 cm^{-1}, which is characteristic of benzene ring. They also have medium sharp absorptions ~3300 cm^{-1}, characteristic of N–H bonds. Spectrum (5) exhibits two such absorptions and, hence, there is an –NH_2 group, while spectrum (6) has only one apparently due to N-H structure. Therefore, we can assign formula C to the spectrum (5) and the formula E to the spectrum (6). Now, we are left with spectra (7) and (8) and formulas A and H. They are easy to distinguish. Formula A has a carbonyl group and is consistent with the spectrum (8), while formula H has an –OH group and in consistent with the spectrum (7). Therefore, the final assignment is:

G-(3); B-(2); F-(4); D-(1); C-(5); E-(6); H-(7) and A-(8).

12. When solving integrated spectroscopy problems like this (those that involve different types of spectra of a compound), it is a good idea to start with the simpler spectra – IR and MS. IR spectrum gives us information about the functional groups and MS about the molar mass. MS may also provide some information about additional structural elements based on the fragmentation pattern. In this case IR absorption at ~3300 cm^{-1} indicates presence of an –OH group. MS spectrum has a molecular ion peak at 88 (do not miss it because it is so small – alcohols frequently give very small molecular ions, or sometimes none at all) and we can assume that the molar mass is 88. Loss of fragment with a mass of 17 (72-55=17) to give the peak at 55, confirms the presence of –OH group. NMR spectra provide much more detailed information. There are five signals in ^{13}C NMR spectrum, which means that there

are five different carbon atoms in the molecule. Furthermore, information from DEPT experiments gives us a number of protons on each carbon. Thus, so far we have the following structural elements:

–OH, two different CH$_2$ groups, one CH and two different CH$_3$ groups.

The total mass of all those groups is: $17 + 2 \times 14 + 13 + 2 \times 15 = 88$. Therefore, those are all of the structural element present in the molecule. To put them together into a structural formula, we should use information from the ^{1}H NMR spectrum. Integration of the spectrum is provided in the form of number of protons above the each signal. Signals due to the two methyl groups are a doublet, which indicated that the methyl group is next to a CH, and a triplet, which indicates that the methyl group is next to a CH$_2$. Therefore, now we have the following structural elements:

CH$_3$CH$_2$– , –OH, –CH$_2$– , –CHCH$_3$.

The only way to put them together into a structural formula is in a form of 2-pentanol and that is the correct answer:

13. By following the procedure outlined above, we can identify –OH group (~3300 cm^{-1}) and a benzene ring (several weak absorptions 2000-1600 cm^{-1}) in the IR spectrum. MS spectrum indicates a molecular ion at 108 (there are also M+1 and M+2 peaks at 109 and 110, respectively). Signals in the MS at 77, 78 and 79 confirm presence of a benzene ring. There are 7 signals for carbons in ^{13}C NMR spectrum, which means that there are seven different carbon atoms in the molecule. They are two C, four CH and a CH$_3$. Total mass of those fragments, combined with –OH which was identified earlier, is 108. Therefore, those are all of the structural elements and now they should be assembled into a structural formula. Presence of two aromatic C and four aromatic CH structures indicates a disubstituted benzene ring, with –OH and –CH$_3$ substituents. Thus, there are three possible structural formulas:

ortho   meta   para

We should use information from the ^{1}H NMR to identify the correct isomer. Signals due to aromatic protons should reveal the substitution pattern. Meta isomer would give a singlet, due to proton on the carbon between the two substituents, two different doublets and one doublet of doublets. This is not the observed pattern and meta isomer can be eliminated. Para isomer has only two different protons and would be expected to give only two signals, each being a doublet. This was not observer either and that one can be eliminated, too. The ortho isomer has four different hydrogens and would be expected to give four different signals two doublets and two multiplets (doublets of doublets) – which is consistent with the spectrum. Note that the multiplets appear close together and have been integrated together as two protons. Therefore, ortho isomer is the correct answer. This particular compound has a trivial name of cresol and the correct answer is ortho-cresol.

Additional Exercises: Questions 14-25.

14. What is the product of the following reaction?

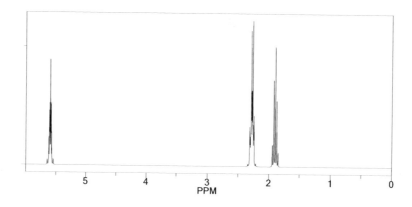

Proton NMR Spectrum of the compound B:

15. The following is a ¹H NMR spectrum of:

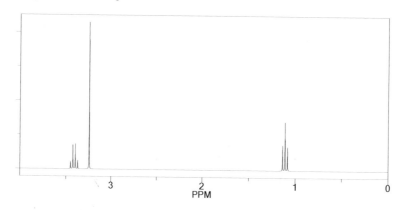

a) 1-propanol

b) 2-propanol

c) ethyl methyl ether

d) ⬜−O

e) ⬜−NH

16. Provide a structural formula of the following compound.

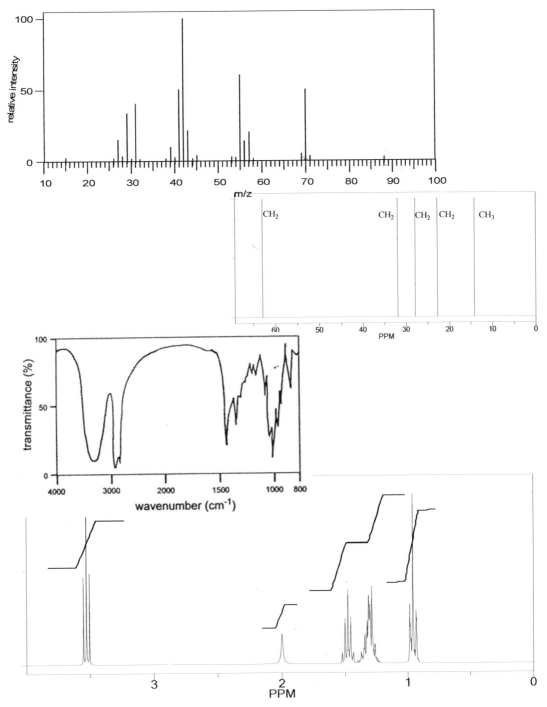

17. Provide a structural formula of the following compound.

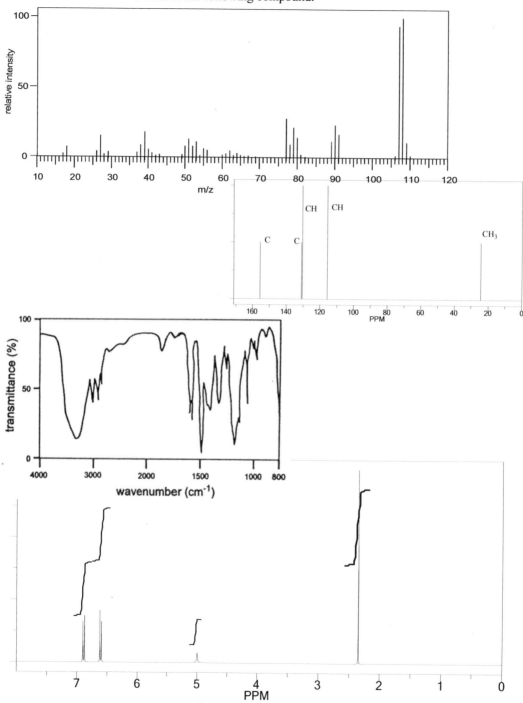

18. IR spectroscopy provides information about:

    a) molar mass.
    b) atoms present in the molecule.
    c) functional groups present in the molecule.
    d) conjugated double bonds.

19. Which spectroscopic technique does not involve interaction of a sample with electromagnetic radiation?

    a) NMR
    b) MS
    c) IR
    d) UV-Vis

20. Mass spectrometry can provide information about:

    a) molar mass
    b) molecular formula
    c) structure
    d) all of the above

21. In UV-Vis spectroscopy:

    a) an electron is promoted to a higher energy level
    b) an electron is ejected from a molecule to give the corresponding cation
    c) spin of a nucleus undergoes change
    d) vibrational and rotational energy levels of the individual bonds are excited

22. The standard used in NMR spectroscopy is:

    a) trimethylsilane
    b) tetramethylsilane
    c) cyclohexane
    d) water

23. Absorption by a carbonyl group in an IR spectrum is:

    a) weak and broad
    b) strong and broad
    c) weak and sharp
    d) strong and sharp

24. Signals in ^{1}H NMR spectrum may be split into multiplets due to:

    a) spin-spin coupling between hydrogen nuclei
    b) spin-spin coupling between hydrogen and carbon nuclei
    c) shielding by electron withdrawing substituents
    d) shielding by electron donating substituents

25. The most intense signal in a mass spectrum is:

   a) fragment ion
   b) molecular ion
   c) parent ion
   d) base peak

Answers to Additional Exercises: Questions 14-25.

14. cyclopentene. 15. c). 16. 1-pentanol.

17. para-cresol.

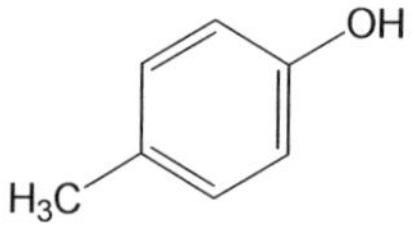

18. c); 19. b); 20. d); 21. a); 22. b); 23. d); 24. a); 25. d).

Exercises in Organic Chemistry

Test 4: Chapters 13-15

1. Give the mechanism and predict the product of the following reaction:

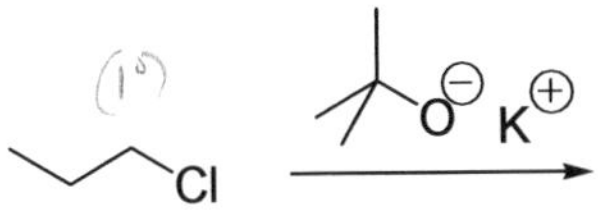

   CH₃CH₂CH₂Cl (1°) + (CH₃)₃CO⁻ K⁺ →

2. How would you prepare the following compound.

   (Br)(H)C=C(H)(H)   from   (H)(H)C=C(H)(H)

3. Provide the major product for each of the following elimination reactions.

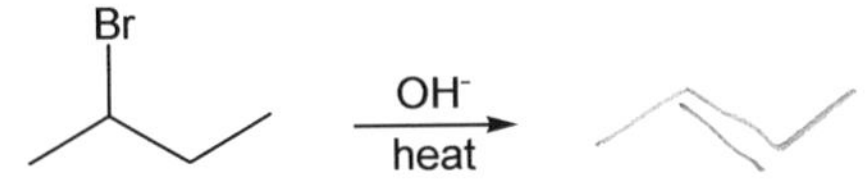

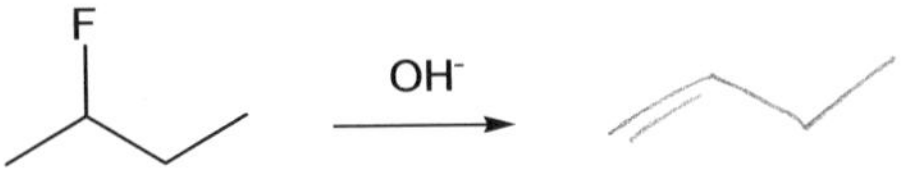

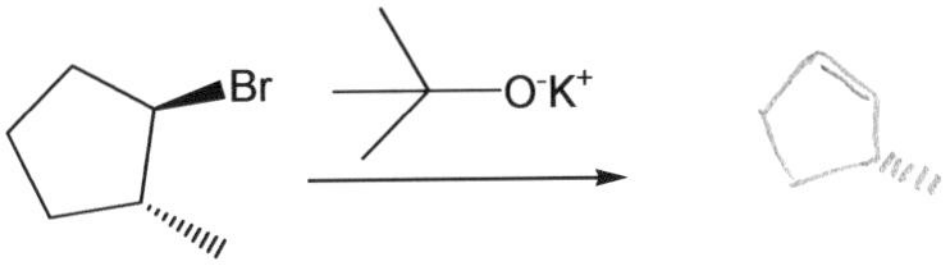

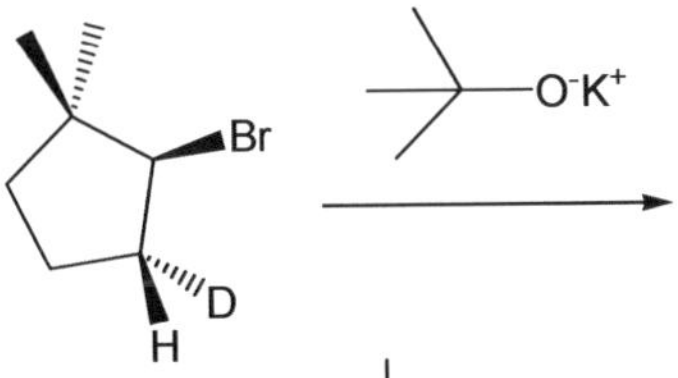

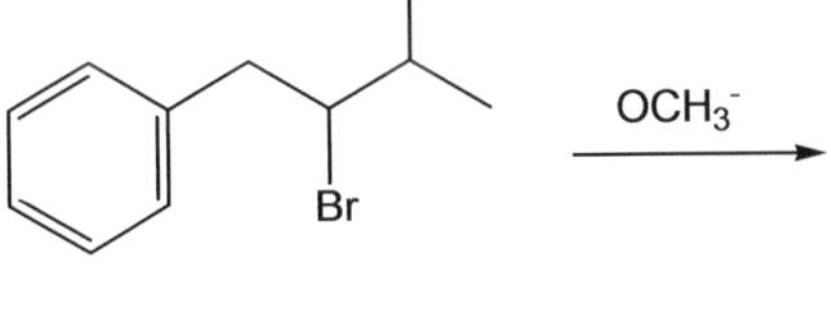

   2-bromobutane + OH⁻, heat →

   2-fluoro-3-methylbutane + OH⁻ →

   trans-1,2-disubstituted cyclopentane (Br, ...) + (CH₃)₃C-O⁻K⁺ →

   cyclopentane with CH₃, Br, D, H + (CH₃)₃C-O⁻K⁺ →

   PhCH₂CH(Br)CH(CH₃)₂ + ⁻OCH₃ →

163

4. Provide the major product for each of the following reactions.

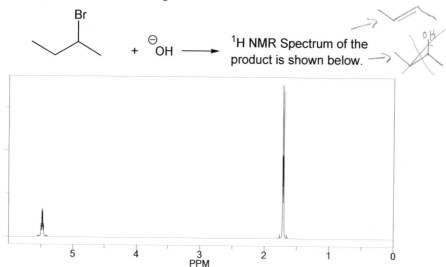

5. Give the mechanism and predict the product of the following reaction:

6. What is the product of the following reaction?

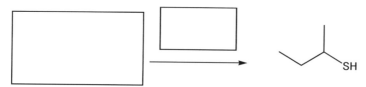

7. For the following reactions provide the missing compounds:

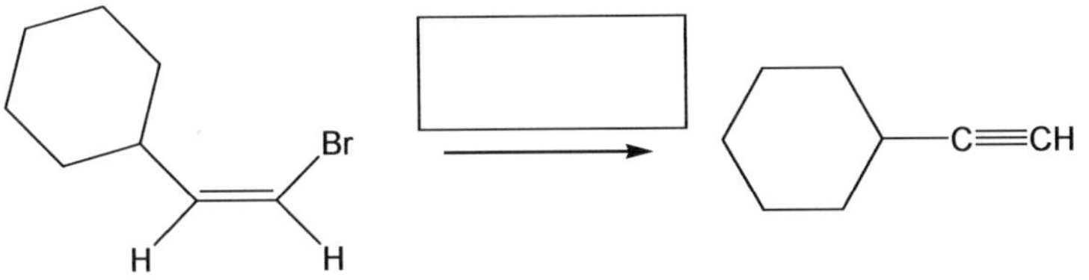

Exercises in Organic Chemistry

PhCH₂OH ⟶ [ ] ⟶ PhCOOH

Cyclohexyl-CH=CHBr ⟶ [ ] ⟶ Cyclohexyl-C≡CH

8. Provide mechanism and final products for the following reactions:

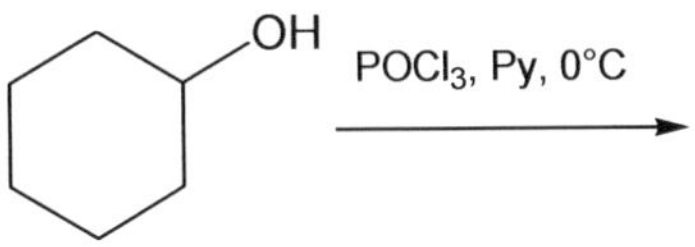

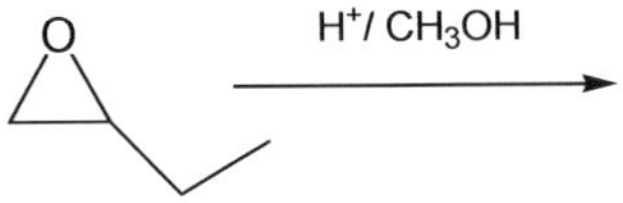

Cyclohexanol (with OH) + POCl₃, Py, 0°C ⟶

Propylene oxide + H⁺ / CH₃OH ⟶

9. A proton NMR spectrum of diethyl ether has:

a) two peaks - a doublet and a triplet.
b) two peaks - a triplet and a quartet
c) four peaks - all doublets
d) four peaks – all triplets

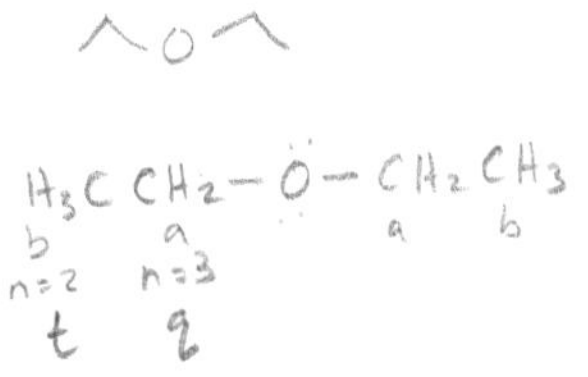

$H_3C\ CH_2-O-CH_2\ CH_3$
  b   a   a   b
 n=2 n=3
  t   q

165

10. A compound shows three signals at in the NMR spectrum at δ = 2.20 (s), 5.08 (s), 7.25 ppm (m). Which compound would satisfy these data?

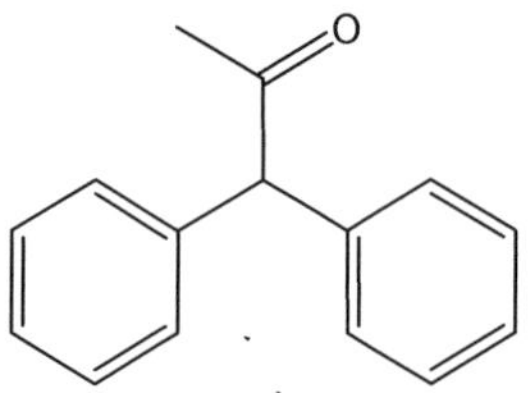

a)

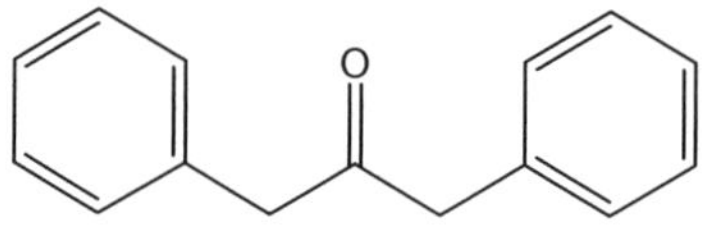

b)

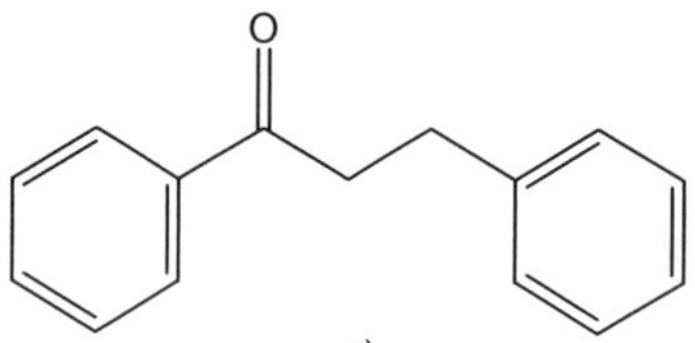

c)

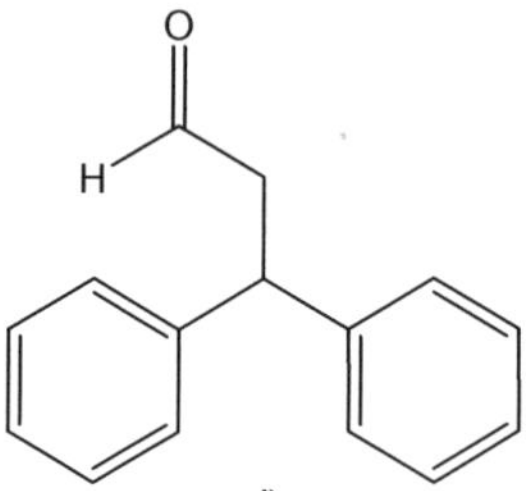

d)

11. How would you follow the progress of the following reaction by means of IR spectroscopy?

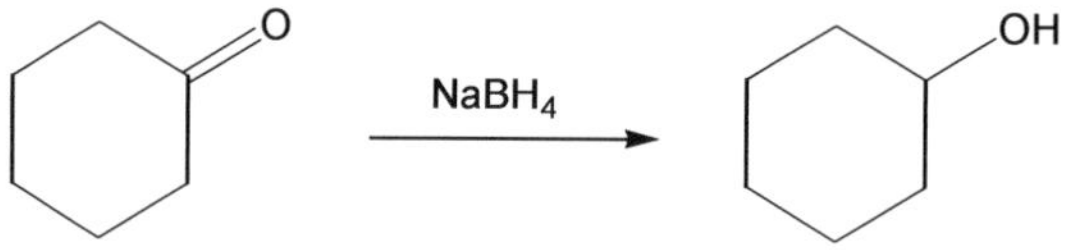

$$\text{cyclohexanone} \xrightarrow{NaBH_4} \text{cyclohexanol}$$

12. How would you follow the progress of the following reaction by means of ^{1}H NMR spectroscopy?

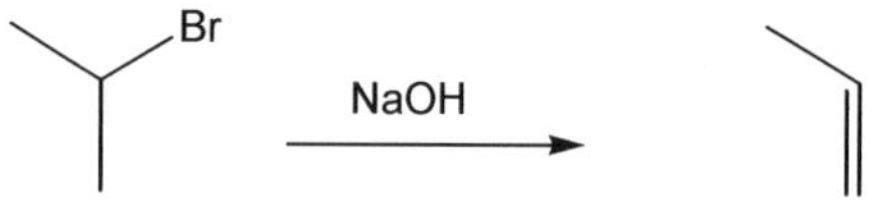

$$\xrightarrow{NaOH}$$

Exercises in Organic Chemistry

13. Provide a structure for a compound that is consistent with the following spectra.

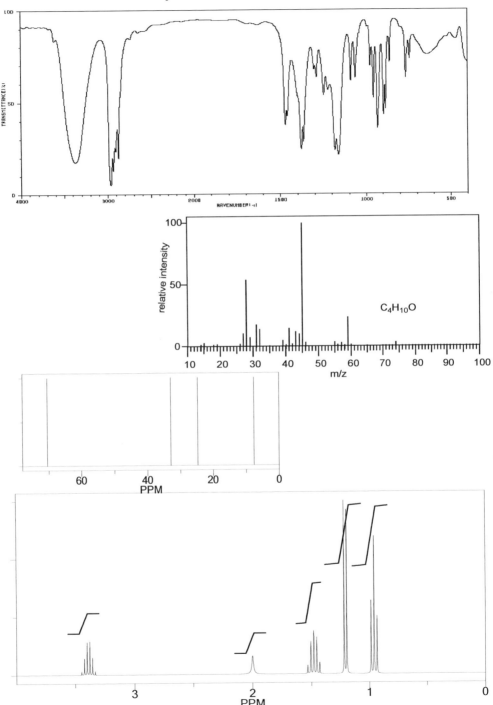

Test 4 Solutions

1. The mechanism is E2 because the elimination is on a primary alkyl halide.

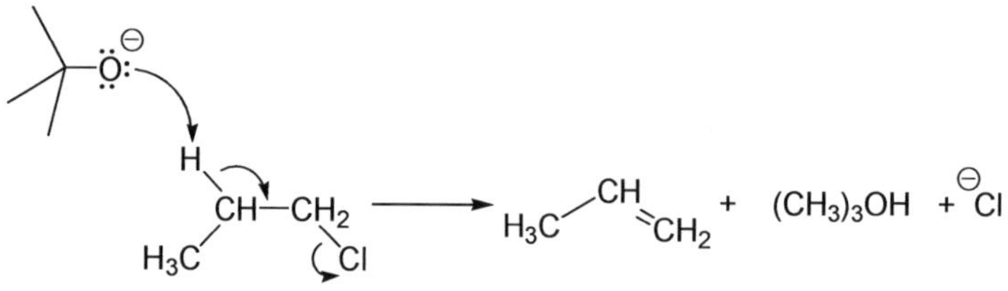
H
CH—CH2
H3C
Cl
H3C
CH
CH2
+
(CH3)3OH
+
Cl

2. Preparation of bromoethene:

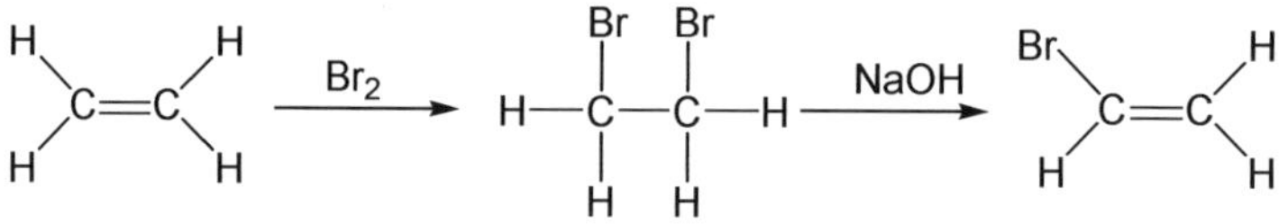
Br2
NaOH

3. Products of elimination reactions:

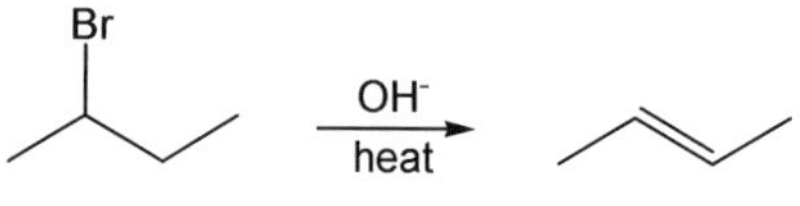
Br
OH-
heat

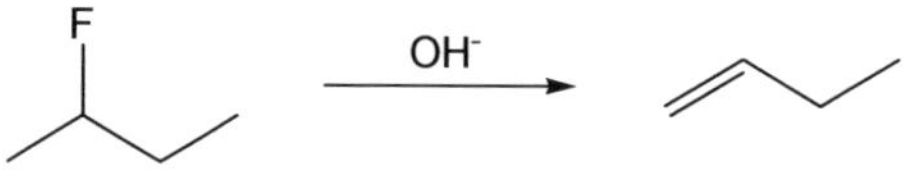
F
OH-

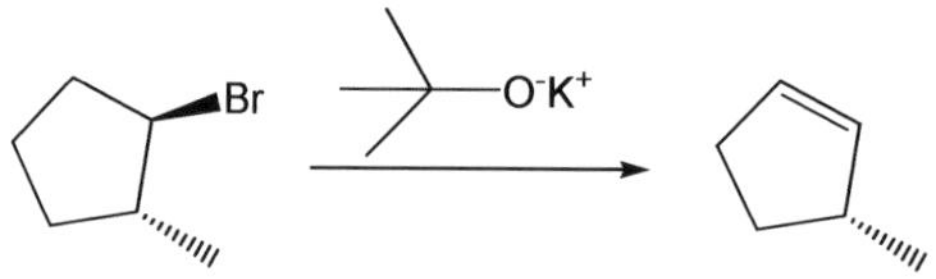
Br
O-K+

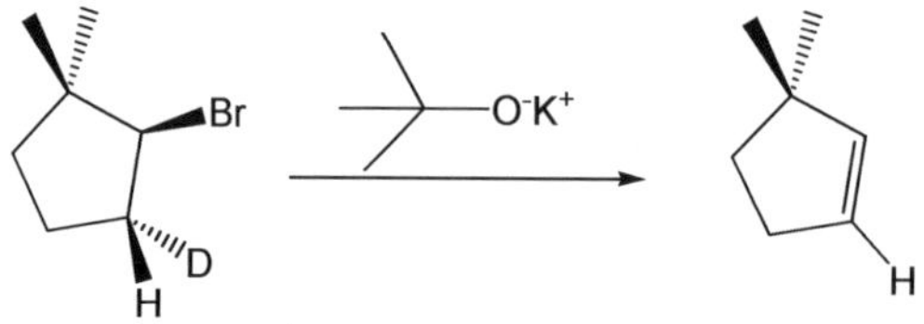
Br
D
H
O-K+
H

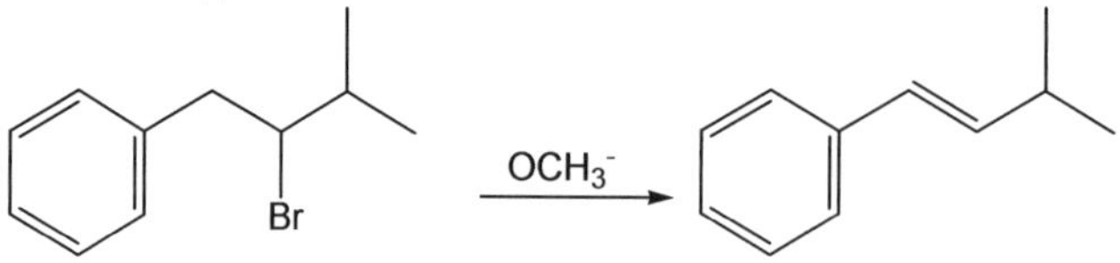
Br
OCH3-

4. Provide the major product for each of the following reactions.

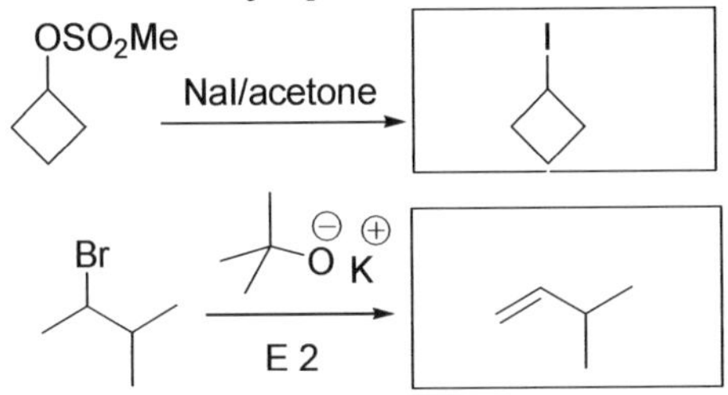

5. The reaction mechanism:

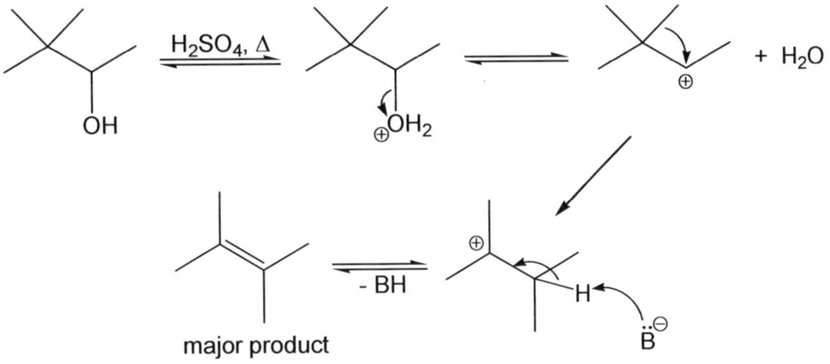

6. Reaction product is 2-butene.

7. The missing compounds:

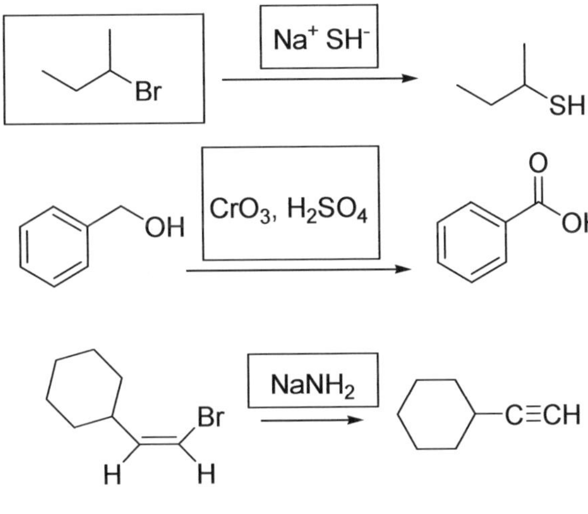

8. Reaction mechanisms:

9. b). 10. a).

11. The starting material exhibits a strong absorption in IR spectrum at ~ 1850-1650 cm^{-1} due to the carbonyl group. As the reaction progresses, this absorption is going to diminish in intensity, while the absorption ~3300 cm^{-1}, due to O-H bond in the product, is going to appear and will grow more intense. The reaction is completed when the absorption at ~ 1850-1650 cm^{-1}, due to the starting material, completely disappears.

12. In ^{1}H NMR spectrum, the starting material has a doublet at ~ 1-1.5 ppm due to the two methyl groups and a multiplet (a septet) due to C-H. As the reaction progresses, these signals are going to diminish in intensity, while the signals ~5-8 ppm, due to alkene hydrogens in the product, are going to appear and will grow more intense. The reaction is completed when the signals due to the starting material completely disappear.

13. IR Spectrum indicates presence of an –OH group. Therefore, the compound is an alcohol. The compounds has molecular formula of $C_4H_{10}O$ (from HRMS). Therefore, its molar mass is 74. LRMS shows a small peak at 74 and larger peaks at 59 (M-15) and 45 (M-29). The two peaks are due to a loss of –CH$_3$ and –CH$_2$CH$_3$, respectively. Therefore, the compound has an ethyl group. From MS, it is not clear whether –CH$_3$ is a part of the ethyl group, or is another methyl elsewhere in the molecule. From the ^{13}C NMR spectrum one can conclude that there are four different carbon atoms. It may be risky to draw any additional conclusions from the ^{13}C NMR spectrum

as signals due primary and secondary carbons show some overlap. However, relatively high downfield shift of –OH bonded carbon (~70 ppm) indicates that it may be a secondary carbon atom. ^{1}H NMR spectrum shows five signals. Signal at 2 ppm is due to –OH group. Signal at 0.95 ppm has integration of 3 and is a terminal methyl group. It is a triplet and, therefore it is bonded to a –CH$_2$- group. That is the ethyl group which was has also been identified from MS spectrum. Signal at ~1.35 ppm is a doublet and also has an integration of 3. Therefore, it is another methyl group and that one is next to a CH. Since there are only four carbon atoms, one can now establish a skeleton of the molecule: CH$_3$CHCH$_2$CH$_3$. In this structure carbon 2 has only three bonds. Therefore, that is where –OH is bonded to and the compound is *2-butanol*. One can use multiplicity an integration of the signals at ~1.45 ppm (1H, quintet) and ~3.4 ppm (1H, sextet) to confirm this structure.

Veljko Dragojlovic

Exercises in Organic Chemistry

16. Electron Delocalization and Resonance

Topics to review: localized and delocalized electrons, resonance structures, resonance energy, molecular orbital description of delocalized electrons, allylic and benzylic cations and radicals, regioselectivity of an electrophilic addition to an alkene, reactivity of substituted alkenes towards electrophiles, inductive and resonance effects of various substituents, electron delocalization and acidity.

Questions 1-5.

1. Which of the following compounds contain delocalized electrons?

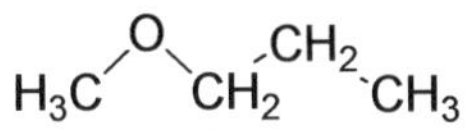

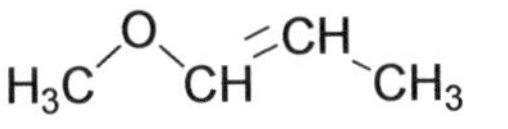

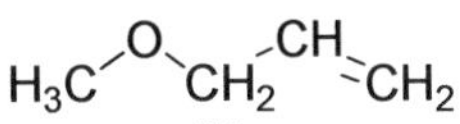

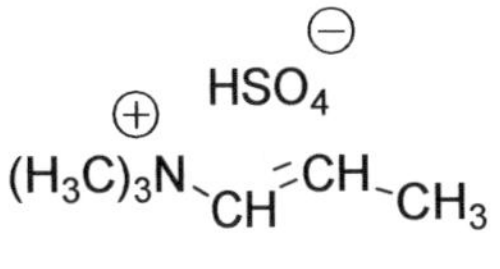

2. Indicate electron flow by means of curved arrows:

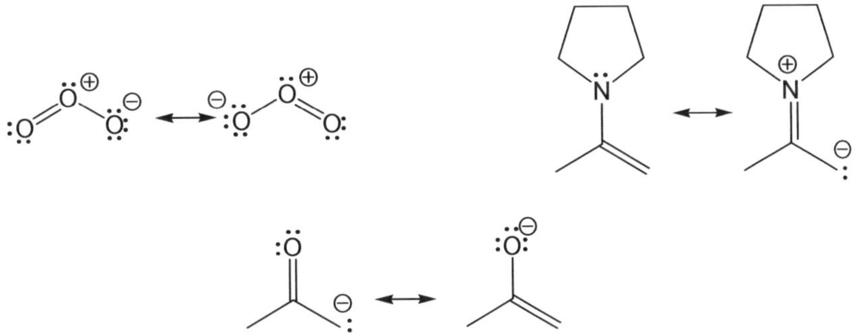

3. Provide the resonance structures indicated by the electron flow (curved arrows). Show formal charges, if any.

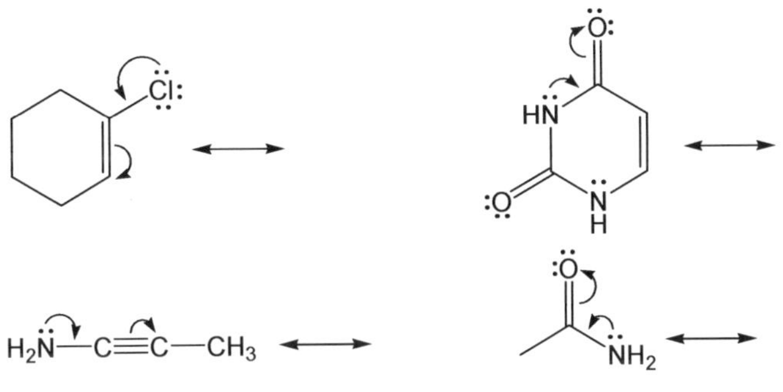

173

4. Show resonance stabilization of the following species by providing curved arrows and resonance structures.

5. In each of the following examples identify the lowest and the highest energy resonance structure.

Answers to questions 1-5.

1. Only compounds II, VI and VII contain delocalized electrons. Each of those compounds has an atom with a free electron pair next to the double bond. In the remaining compounds, either all atoms are sp^3 hybridized, or delocalization is prevented by an sp^3 hybridized atom between a free electron pair and a double bond.

2. Curved arrows:

3. Resonance structures:

4. Resonance stabilization:

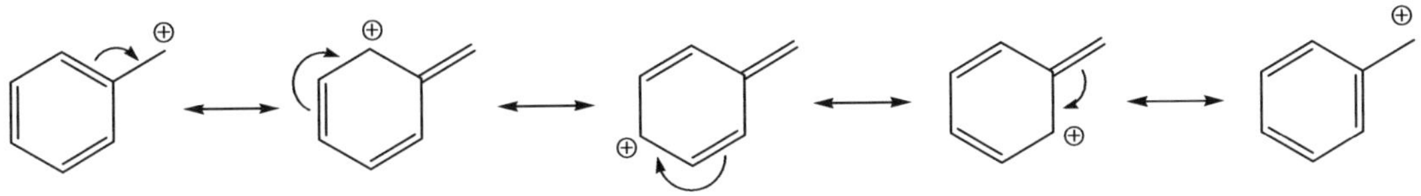

:Br
:Br

H₂C=N=N:
H₂C—N≡N:

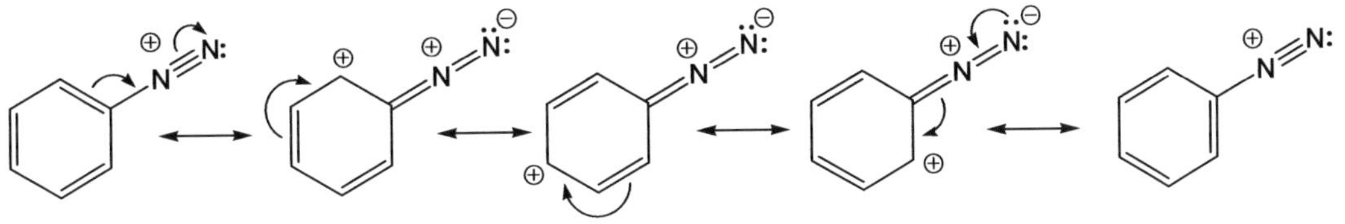
N
N:

5. Energies of resonance structures:

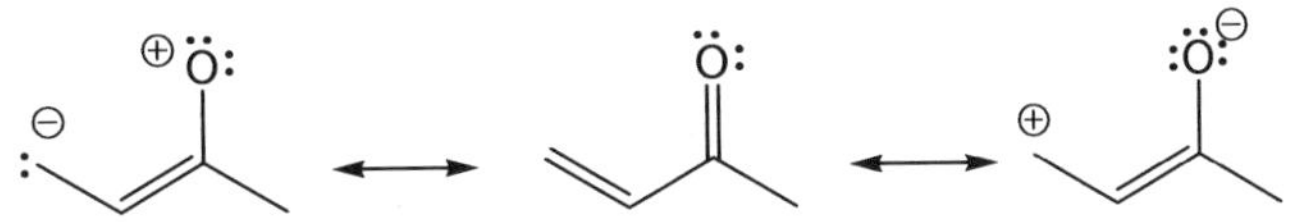
O:
O:
:O:

highest energy　　lowest energy

:F
:F

lower energy　　higher energy

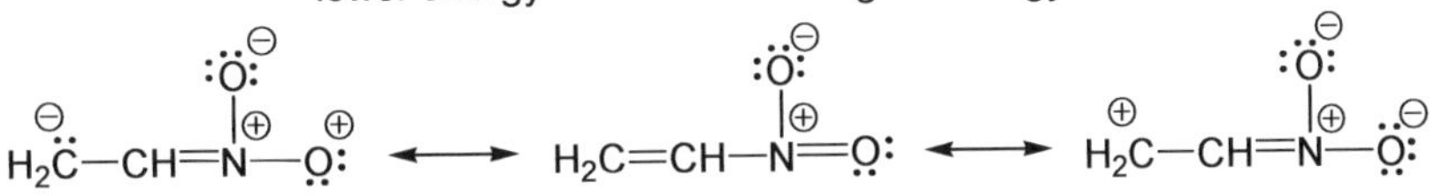
H₂C—CH=N—O:
H₂C=CH—N=O:
H₂C—CH=N—O:

highest energy　　lowest energy

Questions 6 and 7.

6. By means of resonance structures show electron delocalization in 1,3-butadiene.

7. Provide a drawing of relative energy levels of molecular orbitals of allyl cation ($C_3H_5^+$).

Answers to questions 6 and 7.

6. Resonance structures of 1,3-butadiene:

$$\overset{\ominus\,..}{H_2C}-CH=CH-\overset{\oplus}{C}H_2 \longleftrightarrow H_2C=CH-CH=CH_2 \longleftrightarrow \overset{\oplus}{H_2C}-CH=CH-\overset{\ominus\,..}{C}H_2$$

7. Allyl cation has three sp^2 hybridized carbon atoms. According to Molecular orbital theory, each of the three atoms contributes one p orbital to form three molecular orbitals. The lowest energy orbital envelopes the entire molecule and has no nodes. This is a bonding orbital (π). Next orbital has a node in the middle (on the middle carbon atom). It does not contribute to overall bonding and is called a nonbonding orbital (label n). The highest energy orbital has two nodes (between carbons 1 and 2 and carbons 2 and 3) and is antibonding orbital (π^*). The three molecular orbitals are filled according to Hund's rule – from the lowest energy to the highest. Allyl carbocation has only two electrons from its original atomic p orbitals and both are in the lowest energy bonding π orbital.

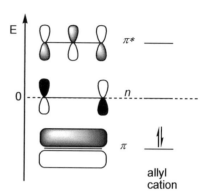

allyl cation

Questions 8-10.

8. Draw a structural formula of the major product of each of the following reactions. Show correct regiochemistry of the product.

   (furan) + HBr, CH$_2$Cl$_2$ →

   (dihydropyran) + HBr, CH$_2$Cl$_2$ →

   $H_2C=CH-C\equiv N$ + HCl →

   $H_3C-CH=CH-NO_2$ + HI →

9. Which of the two reactions is faster and why?

   I) (ethylbenzene) + Br$_2$, hv   or   II) (ethylcyclohexane) + Br$_2$, hv

10. Which of the following alkenes would you expect to react the fastest with water in presence of sulfuric acid?

   a) $CH_2=CH_2$
   b) $(CH_3)_2C=CH_2$
   c) $FCH=CHNO_2$
   d) $CCl_2=CCl_2$

Answers to questions 8-10.

8. Reaction products:

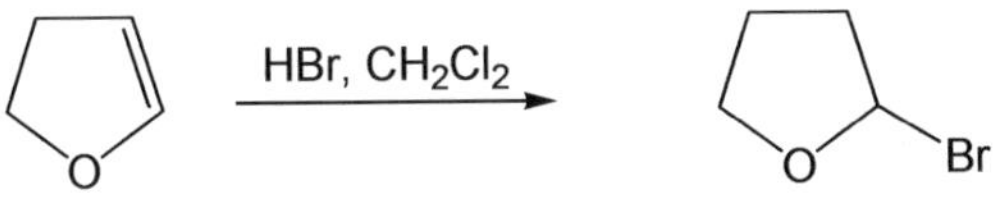

[Furan] $\xrightarrow{HBr, CH_2Cl_2}$ [2-bromotetrahydrofuran] +R effect of oxygen stabilizes neighboring carbocation - nucleophile Br adds next to the oxygen atom

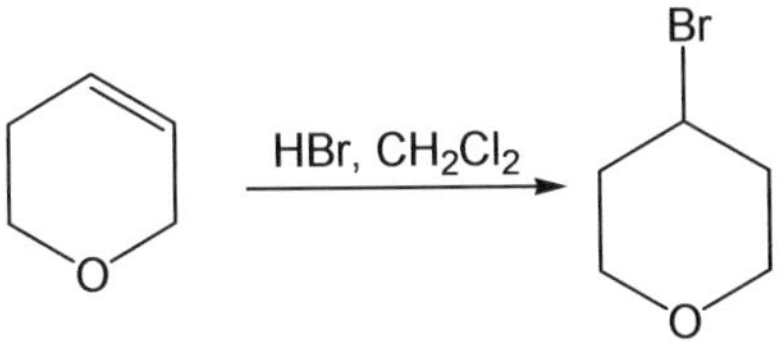

[3,4-dihydro-2H-pyran] $\xrightarrow{HBr, CH_2Cl_2}$ [4-bromotetrahydropyran] Resonance effect is not possible - there is sp^3 carbon between oxygen and double bond. -I effect of oxygen destabilizes carbocation - nucleophile adds away from the oxygen atom.

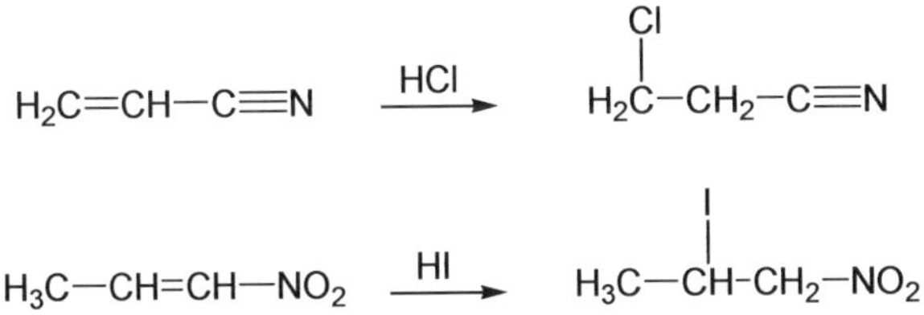

$H_2C=CH-C\equiv N \xrightarrow{HCl} H_2C(Cl)-CH_2-C\equiv N$

$H_3C-CH=CH-NO_2 \xrightarrow{HI} H_3C-CH(I)-CH_2-NO_2$

-R effects of $-C\equiv N$ and $-NO_2$ destabilize neighboring carbocation. Nucleophile adds away from the substituent.

9. Reaction **I** will be faster. The reaction intermediate is benzylic free radical which is more stable (forms faster) compared to free radicals formed from ethylcyclohexane. Note that the regioselectivity of the reactions is different. In case of the reaction **I**, a secondary benzyl radical is the intermediate, while in the reaction **II** it is a tertiary radical.

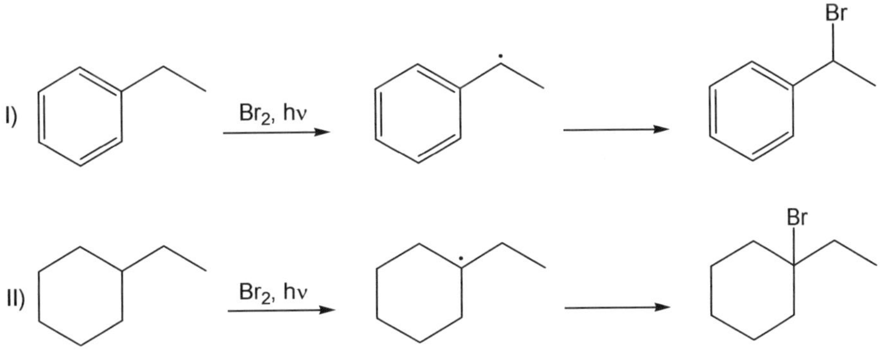

I) [ethylbenzene] $\xrightarrow{Br_2, hv}$ [benzylic radical] $\longrightarrow$ [1-bromo-1-phenylethane]

II) [ethylcyclohexane] $\xrightarrow{Br_2, hv}$ [tertiary radical] $\longrightarrow$ [1-bromo-1-ethylcyclohexane]

10. The rate determining step in this reaction is addition of electrophile (in this case proton) to the double bond. The double bond that is the strongest nucleophile will react the fastest and, in turn, that is the double bond with electron donating groups. Only compound b) has electron donating groups (two methyl groups with their +I effects) and that is the correct answer. Compound a) has no substituents, and c) and d) have electron withdrawing substituents (-F, -NO$_2$ and –Cl).

Questions 11-13.

11. Which of the following gives the correct order of stability of the following cations from the most stable to the least stable?

    a) I>III>II>IV
    b) IV>II>I>III
    c) I>III>IV>II
    d) III>I>II>IV

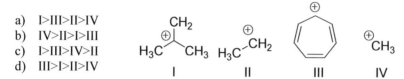

12. Which of the following gives the correct order of stability of the following carbon free radicals, from the most stable to the least stable?

    a) III>II>I>IV
    b) III>II>IV>I
    c) II>III>I>IV
    d) II>I>III>IV

13. What is the sequence of reactivity towards hydrogen chloride of the following alkenes?

    I. FCH=CH$_2$ II. (CH$_3$)$_2$C=CH$_2$ III. CH$_3$OCH=CH$_2$ IV. (NC)$_2$C=C(CN)$_2$

    a) III>II>I>IV
    b) II>III>I>IV
    c) IV>I>II>III
    d) I>IV>III>II

Answers to questions 11-13.

11. Among the four carbocations, one is stabilized by resonance and it is the most stable one. That is carbocation III (cycloheptatrienyl carbocation). The remaining carbocations are stabilized through hyperconjugation and their stability increases with increasing degree of substitution. Therefore, tertiary (I) is the most stable of the three and methyl (IV) the least and the overall order is: III>I>II>IV or d).

12. Among the four free radicals, one is stabilized by resonance and it is the most stable one. That is benzyl free radical II. The stability of remaining free radicals increases with increasing degree of substitution. Therefore, tertiary (III) is the most stable of the three and methyl (IV) the least and the overall order is: II>III>I>IV or c).

13. As was the case with the question 10, the rate determining step in this reaction is addition of electrophile (proton) to the double bond. The double bond that is the strongest nucleophile reacts the fastest and, in turn, that is the double bond with electron donating groups. Compound **I** has –F substituent with its –I effect, which decreases reactivity of the double bond. Compound **II** has two methyl groups (with +I effects) which increase reactivity of the double bond. Therefore **II** > **I**. Compound **III** has a methoxy substituent with +R effect. Resonance effect is stronger than inductive effect and this compound will be even more reactive than **II**. Thus, so far the order of reactivities is **III** > **II** > **I**. Finally compound **IV** has four –CN groups which have –R effects. This is a highly electron deficient alkene is by far the least reactive of the four. The overall order of reactivates is **III** > **II** > **I** > **IV** or a).

Additional Exercises: Questions 14-20.

14. Which of the following ions are stabilized by delocalization of electrons?

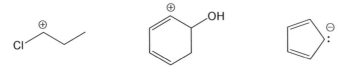

15. Provide a drawing of relative energy levels of molecular orbitals of 1,3-butadiene.

16. The order of reactivity of the following alkenes towards an electrophilic attack is:

   I. $F_2C=CF_2$ II. $CH_3SCH=CH_2$ III. $CH_3CH=CHCH_3$ IV. $Br_2C=CH_2$

   a) III>II>I>IV
   b) II>III>IV>I
   c) IV>I>II>III
   d) I>IV>III>II

17. Which of the following carbocations is the most stable?

   a) Ph_3C^+
   b) $CH_3CH_2^+$
   c) $(CH_3)_2CH^+$
   d) $CH_3CH^+CH_3$

18. Inductive effects of the groups $-OCH_3$, $-Si(CH_3)_3$, $-Br$, $-NH_3^+$ are respectively:

   a) +I,   -I,   +I,   +I
   b) -I,   +I,   -I,   +I
   c) -I,   -I,   +I,   +I
   d) -I,   +I,   -I,   -I

19. Resonance effects of the groups –NH$_2$, -O⁻, -Cl, -N⁺≡N are respectively:

   a) +R,   -R,   +R,   +R
   b) +R,   +R,   +R,   -R
   c) -R,   -R,   -R,   +R
   d) -R,   +R,   -R,   -R

20. Which of the following carbocations is the most stable?

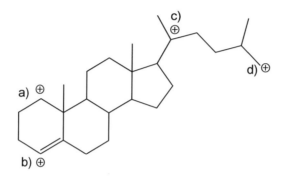

Answers to Additional Exercises: Questions 14-20.

14. All three ions are stabilized through delocalization of electrons.
15. Molecular orbitals of 1,3-butadiene are:

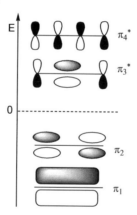

16. b); 17. a); 18. d); 19. b); 20. c).

Exercises in Organic Chemistry

17. Dienes and Polyenes

Topics to review: structure and nomenclature of dienes and polyfunctional compounds, reactivity of isolated and conjugated dienes, kinetic versus thermodynamic control of a reaction, chemoselectivity, Diels – Alder reaction.

Questions 1-6.

1. What is the final product of the following sequence of reactions?

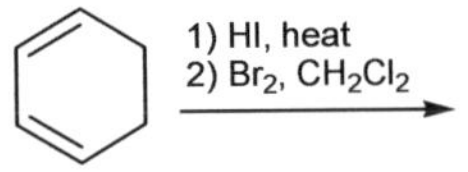

   cyclohexa-1,3-diene
   1) HI, heat
   2) Br$_2$, CH$_2$Cl$_2$

2. What are the products of the following reactions? Show correct regioisomers and stereoisomers, where applicable.

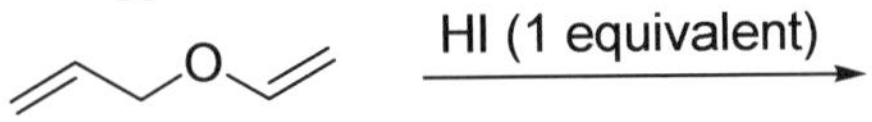

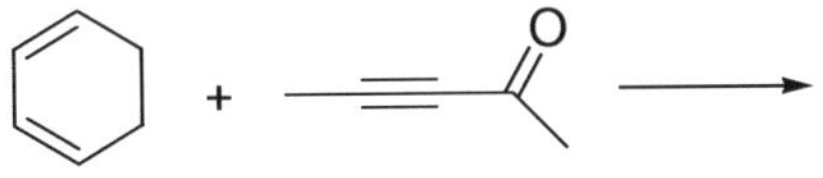

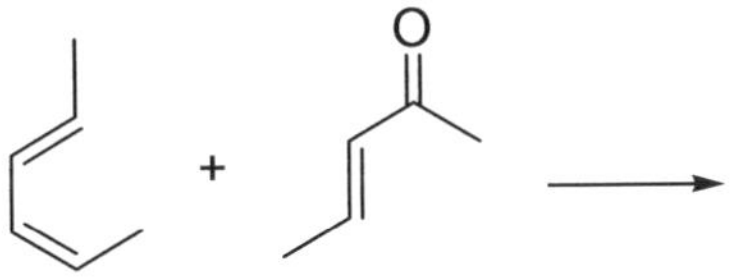

   allyl vinyl ether —HI (1 equivalent)→

   cyclohexene + methyl propynyl ketone →

   (Z,Z)-hexa-2,4-diene + methyl vinyl ketone →

3. How would you prepare the following compound?

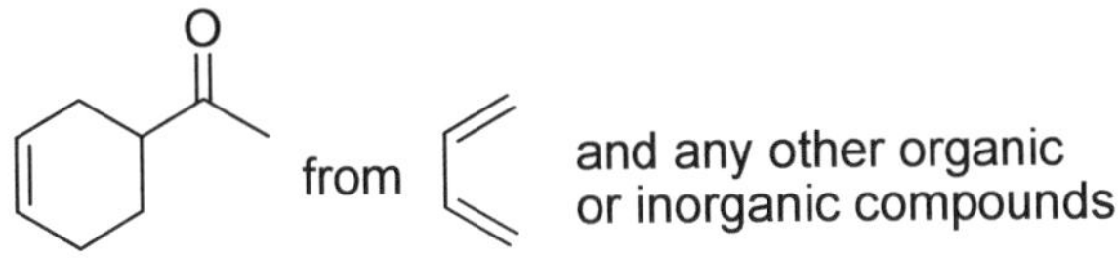

   (1-acetyl-3-cyclohexene) from (1,3-butadiene) and any other organic or inorganic compounds

4. What was the starting material in the following reaction?

   ? —HBr (1 equivalent)→ (2-bromo-2,3-dimethyl-hex-5-ene)

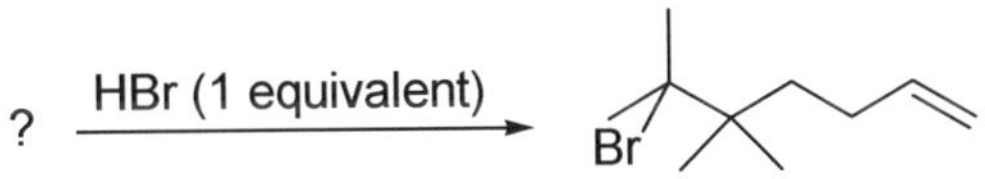

5. Provide the reagent for the following reaction?

   cyclopentadiene —?→ 3,4-dibromocyclopentene

181

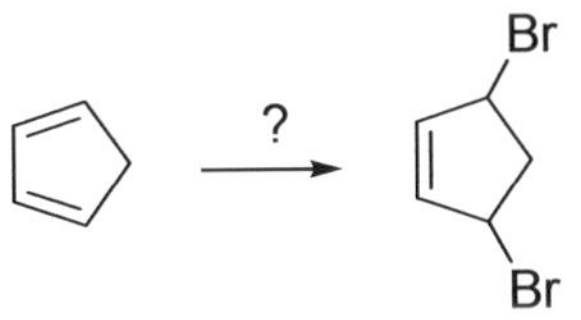

6. Which of the following compounds has the lowest heat of hydrogenation?

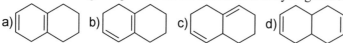

Answers to questions 1-6.

1. The first step of the reaction is addition of HI to diene. The addition is a thermodynamic 1,4-addition since reaction is done under thermodynamic conditions (with heating) to give the following compound:

In the second step, a molecule of bromine adds across the double bond. The most likely stereochemistry of the product is shown below. Due to steric hindrance bromonium ion forms on the side of the ring opposite to the bulky iodine substituent. Bromine nucleophile then adds to the carbon atom away from iodine.

2. Reaction products:

This double bond is deactivated by -I effect of oxygen.

This double bond is activated by +R effect of oxygen.

Note the way these two carbons are represented in the product.

In order to figure out the correct stereochemistry, add hydrogen atoms and assume that the dienophile comes from the top.

3. The product contains a six-member ring and Diels – Alder reaction is one way to prepare it. Since a starting material is a diene, one should find out what is the dienophile. Therefore, one has to start with the product and work it backwards to the starting materials.

Break these bonds to end up with diene and dienophile.

4. Starting material is a diene because the reaction is an addition reaction of HBr and there is one double bond left in the final product. The double bond that reacts, is a more substituted and, hence, more reactive double bond at the other end of the chain.

5. The reagent is bromine, Br_2, and, in 1,4-addition to diene, it gives the reaction product.

6. When comparing alkenes, one with the lowest heat of hydrogenation is the most stable one. Therefore, the actual question is: "Which of the following alkenes is the most stable?" The compound b) is the only conjugated diene of the four. Therefore, this is the most stable compound and b) is the correct answer.

Questions 7 and 8.

7. Which of the following compounds are suitable dienophiles?

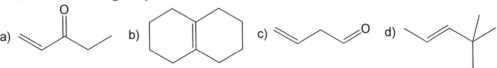

8. Which of the following dienes are suitable substrates for a Diels – Alder reaction?

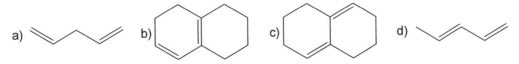

Answers to questions 7 and 8.

7. A dienophile has an electron withdrawing substituent on the carbon – carbon double bond. Only compound a) is a suitable dienophile. It has an electron withdrawing group on the double bond. Compounds b) and d) have electron donating groups on the double bond. Electron withdrawing group of compound c) is at a considerable distance from the double bond and exhibits only a weak –I effect.

8. Only conjugated dienes are suitable substrates in Diels – Alder reactions. Diene a) is not a conjugated diene and can be eliminated. Another requirement is that a conjugated diene must be able to assume a *"cisoid"* conformation so that is can react with a dienophile:

"cisoid" conformation   "transoid" conformation

   Diene c) has two double bonds that are parts of two separate rings and is not able to assume this conformation. Therefore, it can be eliminated, too. Diene b) is already in a correct conformation and diene d) can assume it by a rotation about a single bond. Therefore, dienes b) and d) are suitable substrates for a Diels – Alder reaction.

Additional Exercises: Questions 9-14.

9. Complete the following table.

Name	Structural formula
(2E,4Z)-2-chloro-2,4-hexadiene	
4,4-dibromo-1-penten-3-ol	

Exercises in Organic Chemistry

10. Predict the major product of the following reactions.

    CH₂=C(CH₃)–C(CH₃)=CH₂ (isoprene-like diene) → HBr (1 equivalent) / diethyl ether

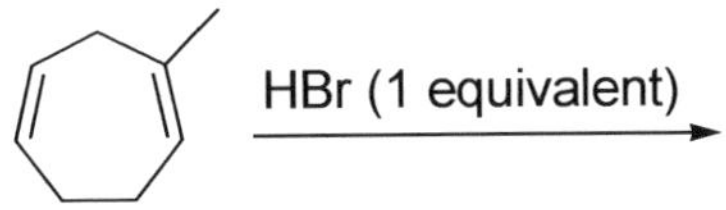

    (1-methylcyclohepta-1,3-diene) → HBr (1 equivalent)

    (butadiene) + NC–C(=CH₂)–CN →

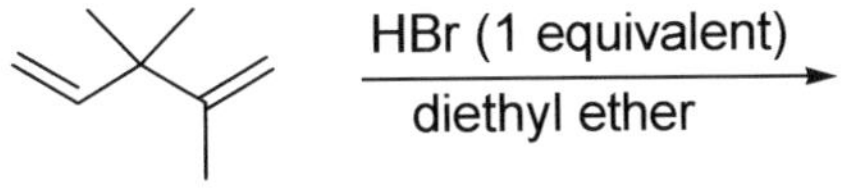

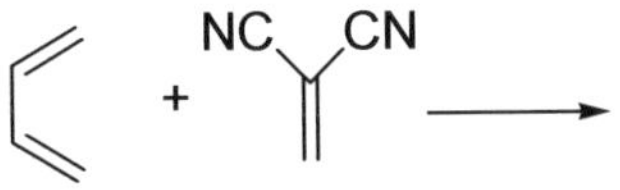

11. Provide the starting material for each of the following reactions.

    ? → Br₂, CH₂Cl₂ → (cyclobutene with two CH₂Br groups)

    ? → Br₂, CH₂Cl₂ → Br–CH₂–C(Ph)(Br)–CH₂–CH=CH₂ (approximate: PhC(Br)(CH₂Br)CH₂CH=CH₂)

    (butadiene) + ? → (cis-4-cyclohexene-1,2-dicarboxylic anhydride)

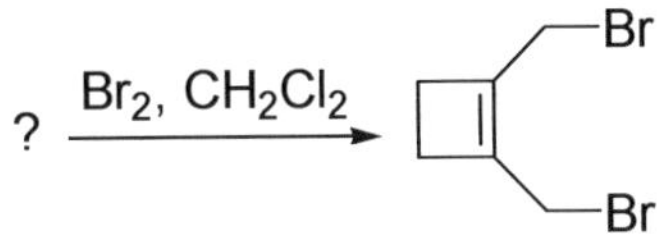

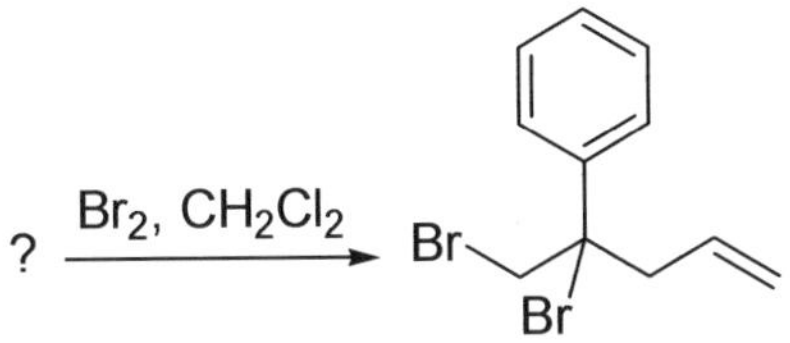

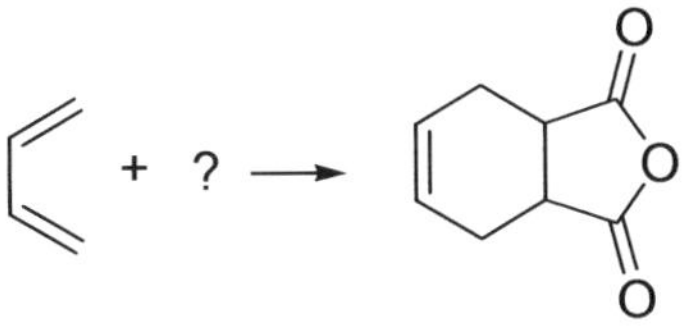

12. Provide the reagent for each of the following reactions.

    (2,6-dimethylocta-2,6-diene type) → ? → (dibromide product with Br at the two former alkene carbons)

    (4-isopropenyl-1-methylcyclohexene / limonene) → 1) ? 2) ? → (product with tertiary Br on ring and tertiary Cl on side chain)

185

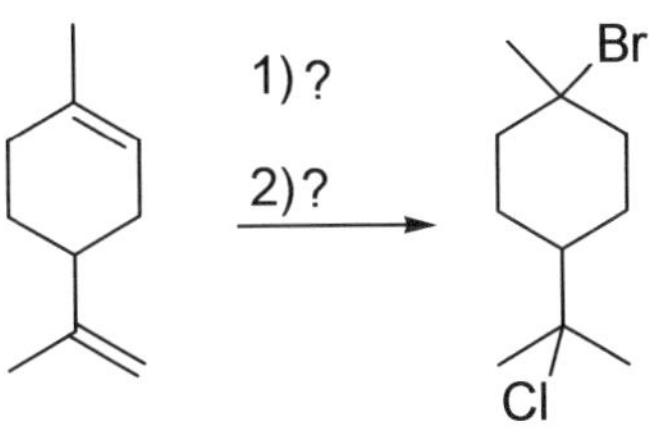

13. For the following reactions provide the formulas of the missing compounds:

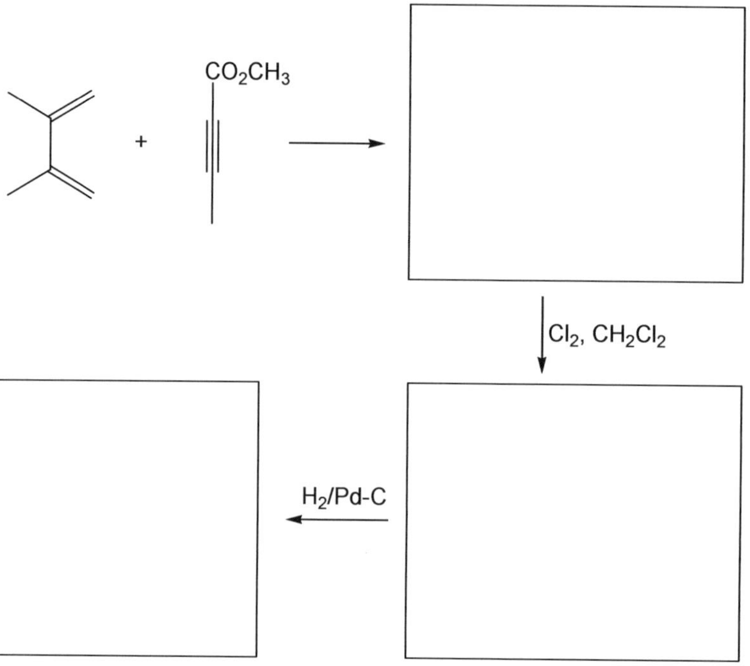

[Reaction scheme: 2,3-dimethyl-1,3-butadiene + methyl propiolate (HC≡C–CO₂CH₃) → (box); then Cl₂, CH₂Cl₂ → (box); then H₂/Pd-C → (box)]

14. Give two sets of diene and dienophile that could be used to prepare the compound below. Which of the two sets is the better one?

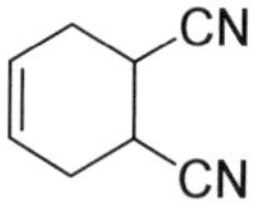

[Structure: cyclohex-4-ene-1,2-dicarbonitrile]

Answers to Additional Exercises: Questions 9-14.

9. Table.

Name	Structural formula
(E)-2,3-dimethyl-2,4-hexadiene	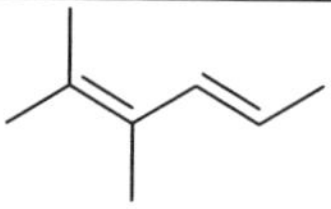 [structure]
(2E,4Z)-2-chloro-2,4-hexadiene	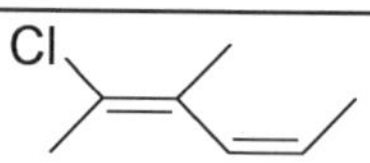 [structure with Cl]
1-penten-4-yne	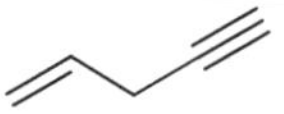 [structure]

4,4-dibromo-1-penten-3-ol	[structure: CH2=CH-CH(OH)-CBr2-CH3]
(E)-2,4-dimethyl-2,4-heptadiene	[structure]
(E)-2-hepten-5-yne	[structure]
1,3-cyclohexadiene	[structure]
(Z)-5-chloro-2-methyl-1,4-heptadiene	[structure]

10. Reaction products.

[structure] + HBr (1 equivalent), diethyl ether → [structure with Br]

[structure] + HBr (1 equivalent) → [structure with Br]

[butadiene] + [NC-C(=CH2)-CN] → [cyclohexene with two CN groups]

11. Starting materials.

[1,2-bis(methylene)cyclobutane] + Br2, CH2Cl2 → [cyclobutene with two CH2Br groups]

[2-phenyl-1,4-pentadiene] + Br2, CH2Cl2 → [product with two Br]

[butadiene] + [maleic anhydride] → [tetrahydrophthalic anhydride]

187

12. Reagents.

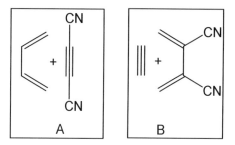

13. Reaction sequence:

14. The two sets, A and B, are shown below. Set A is the better one because it contains a dienophile with two electron withdrawing groups. Ethyne (acetylene) is not a good dienophile.

18. Aromaticity: Reactions of Benzene

Concepts to review: aromaticity, Hückel rule, heterocyclic compounds, electrophilic aromatic substitution, reactions of benzene, nomenclature of phenyl ketones (phenones).

Questions 1-3.

1. Calculate the number of π electrons, apply Hückel's rule, and predict which of the following compounds exhibit aromatic stabilization.

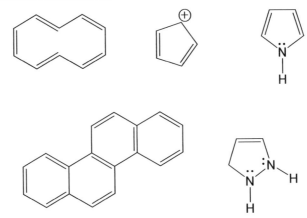

2. Carbene is a high energy neutral carbon species that has two electrons on a divalent carbon atom. A carbene may be either a ***singlet carbene*** in which electrons are paired up and both electrons occupy a hybrid sp^2 orbital, while the atom's p orbital is empty, or a ***triplet carbene***, in which the two electrons are unpaired and each occupies one sp^3 hybrid orbital. Shown here are the two forms of C_3H_2 carbene. For this particular carbene, is the singlet or the triplet form more stable and why? Carbenes usually readily undergo addition, or dimerization, reactions. However, the C_3H_2 carbene is relatively stable (it has been observed in the interstellar space). Why?

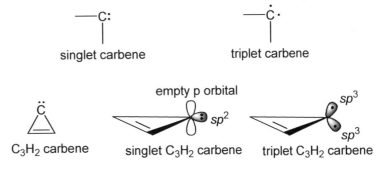

3. Three of the four molecules, or ions, shown below are known and have been observed. So far, chemists have not been able to prepare the fourth one. Identify the molecule, or ion, that has not been prepared. Explain why that molecule, or ion, has not been prepared so far.

Answers to questions 1-3.

1. According to Hückel's rule, planar cyclic compounds with a closed ring of $4n+2$ π electrons exhibit aromatic stabilization.

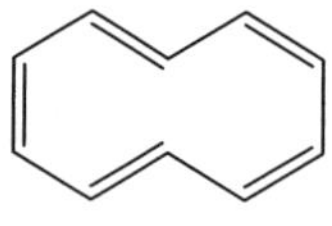

$n = 10$
aromatic

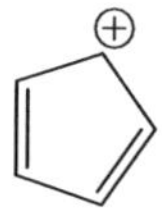

$n = 4$
not aromatic
(antiaromatic)

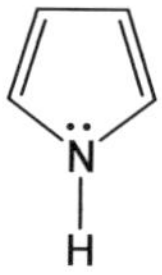

$n = 6$
aromatic

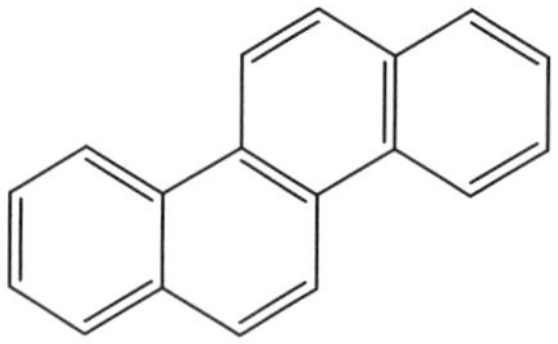

$n = 18$
aromatic

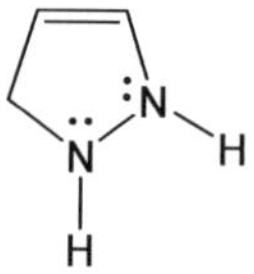

Although $n = 6$, this compound is not aromatic. As has an sp^3 hybridized carbon atom, there is no closed ring of π electrons.

2. In this particular carbene, singlet form is aromatic – it has a closed loop of 2 π electrons. Triplet is a simple cyclic species without aromatic stabilization – it has an sp^3 hybridized carbon atom which prevents electron delocalization throughout the ring. Therefore, in this particular case, singlet carbene is more stable of the two.

3. Cyclic compounds can be aromatic, which are stabilized by electron delocalization, antiaromatic, which are destabilized by electron delocalizaton, and simple cyclic compounds in which there is no closed loop of delocalized electrons (they may have part of a ring with delocalized electrons). Stability of those species is aromatic>simple cyclic>antiaromatic compounds. Of the four species, cation I is aromatic. Molecules II and III are simple cyclic alkenes. They are highly strained due to the presence of a double bond in a three-member ring. Therefore, they are rather unstable, but can be prepared. Molecule IV has four π electrons in a closed loop, which makes it antiaromatic. Therefore, this one is destabilized by electron delocalization. It is by far the least stable of the four and this is the one that has not been prepared as yet.

Questions 4-7.

4. Provide a detailed mechanism for the following reaction. Indicate flow of electrons by means of curved arrows.

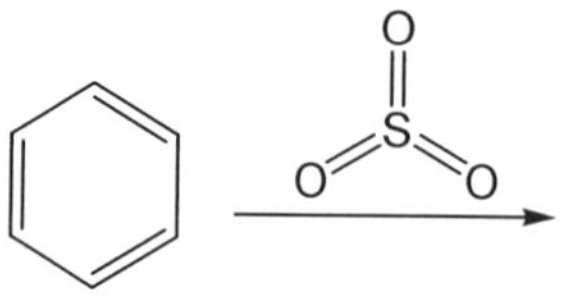

5. Show by means of chemical equations how would you prepare acetophenone starting with benzene and any other organic or inorganic reagents?

Exercises in Organic Chemistry

6. How would you prepare butylbenzene starting with benzene and any other organic or inorganic reagents?

7. Provide formulas of the missing compounds.

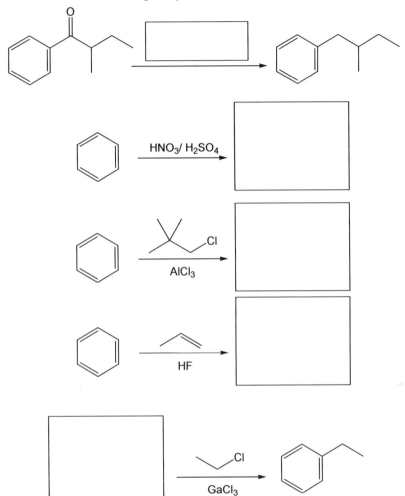

Answers to questions 4-7.

4. A mechanism of sulfonation of benzene is:

An alternative, and probably better, mechanism involves removal of the proton by an intermolecular step followed by protonation of the sulfonate group:

5. Acetophenone is another name for methyl phenyl ketone. It can be prepared in a reaction between benzene and acetyl chloride in the presence of a stochiomeric amount of aluminum trichloride.

6. Butylbenzene is an alkylbenzene. However, a simple straightforward method for preparation of alkylbenzenes, Friedel-Crafts alkylation, would not give the desired product. As a carbocation is the reaction intermediate, the reaction would proceed with a rearrangement and butylbenzene would be only a minor reaction product.

minor   major

In order to avoid a rearrangement, one has to do Friedel-Crafts acylation followed by a reduction of the carbonyl group. Therefore, one has to first react benzene with butanoic acid in the presence of aluminum chloride.

Next, the carbonyl group is reduced to a methylene group either with zinc in hydrochloric acid (Clemmensen reduction) or hydrazine in presence of potassium hydroxide (Wolf–Kishner reduction) to give the desired product.

7. Missing compounds:

Additional Exercises: Questions 8-13.

8. Apply Hückel's rule and predict which of the following compounds exhibits aromatic stabilization.

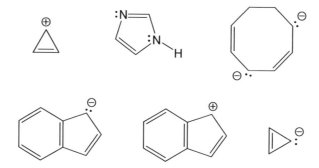

9. Provide formulas of the missing compounds.

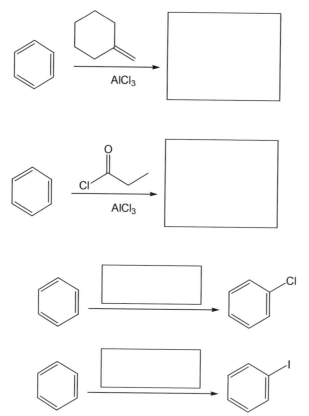

10. The principal reaction of aromatic compounds is:

    a) nucleophilic substitution
    b) electrophilic substitution
    c) nucleophilic addition
    d) electrophilic addition

11. What is hybridization of nitrogen atom in NO_2^+?

    a) sp^3
    b) sp^2
    c) sp
    d) dsp^2

12. An intermediate in a Friedel-Crafts acylation reaction is:

    a) a carbocation.
    b) an acylium cation.
    c) a chloronium ion.
    d) a free radical.

13. Aluminum trichloride in a Friedel-Crafts acylation reaction is:

    a) a catalyst.
    b) an intermediate.
    c) a drying agent.
    d) a stochoimetric reagent.

Answers to Additional Exercises: Questions 8-13.

8. Aromatic compounds:

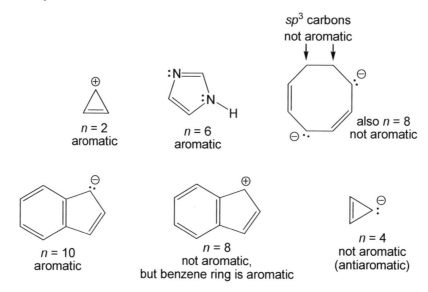

9. Missing compounds:

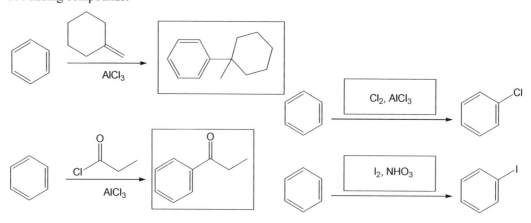

10. b); 11. c); 12. b). 13. d).

19. Reactions of Substituted Benzenes

Concepts to review: nomenclature of substituted benzenes, reactions of substituents on benzene, effect of substituent on reactivity of benzene, synthesis of di- and trisubstituted benzenes, diazonium salts and their reactions, benzyne.

Questions 1-6.

1. Name the following compounds.

2. Which of the following substituents exhibit a strongly deactivating effect on benzene ring?

   OH, $\overset{+}{N}(CH_3)_3$, NO_2, Cl, NH_2

3. Arrange the following compounds in the order of increasing acidity.

4. What is going to be the major product of sulfonation of iodobenzene?

5. What is going to be the major product of nitration of *m*-cresol?

6. How would you prepare benzonitrile starting with benzene and inorganic reagents? Note that any preparation will require several steps.

Answers to questions 1-6.

1. Names of the compounds:

   1,4-dichloro-2-nitrobenzene   3,5-dinitrotoluene   1-chloro-3-fluoro-5-nitrobenzene   2-chloro-3-nitrotoluene

   4-bromo-3-nitrotoluene   1-bromo-2,4-dinitrobenzene   3,5-dichlorotoluene   1-chloro-4-iodo-2-nitrobenzene

2. Common reaction of benzene ring is an electrophilic aromatic substitution reaction. A substituent that exhibits a strong electron withdrawing effect also exhibits a strong deactivating effect on the ring. Those are substituents that have either a positively charged or a highly electron deficient atom bonded to a benzene ring. In the list above, those are ammonium and nitro substituents: —$N(CH_3)_3^+$ —$N^+(=O)O^-$

3. Acid strength of a substituted phenol depends on the electronic effect of the substituent. Electron withdrawing substituents increase acid strength, while electron donating decrease it. Effects of each of the substituents are listed below. o-Nitrophenol, with a strongly electron withdrawing substituent, is the strongest acid, while o-aminophenol, with a strongly electron donating substituent is the weakest of the four acids. The remaining two acids are of the intermediate strength and are arranged in the order of their inductive effects.

   H_2N-C$_6$H$_4$-OH   < H_3C-C$_6$H$_4$-OH   < F-C$_6$H$_4$-OH   < O_2N-C$_6$H$_4$-OH

   +R (major)   +I (only)   −I (major)   −R (major)
   −I (minor)   +R (minor)   −I (minor)

4. Iodine substituent of iodobenzene is weakly deactivating substituent, due to a negative inductive effect of iodine atom, and an ortho, para directing substituent, due to a positive resonance effect of iodine. Thus, the two products will be 4-iodosulfonic acid and 2-iodosulfonic acid. Both −I as well as −SO_3H substituents are rather large and when they are next to each other as in 2-iodosulfonic acid, there will be a considerable steric strain in the molecule. Therefore, 4-iodosulfonic acid is a more stable compound and is the major product of the reaction.

   C$_6$H$_5$I   →[H_2SO_4 (fuming)]   4-iodobenzenesulfonic acid (major)   +   2-iodobenzenesulfonic acid (minor)

5. To answer this question, one has to know what is *o*-cresol. A Cresol is a compound that has both an –OH and –CH$_3$ substituents on a benzene ring. Of course, in *o*-cresol the two substituents are ortho to each other. Next, one has to examine the directing effects of the two substituents. Both of them are activating substituents and have ortho, para-directing effects. Therefore, each substituent will direct the incoming nitro group ortho, para with respect to its own position. Directing effect of –OH group is indicated by asterisks and that of –CH$_3$ by a hollow dot:

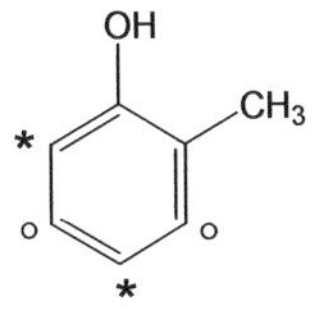

Therefore, the two substituents direct the incoming nitro group into different positions. Between the two substituents, oxygen has a strong activating effect, while methyl group a weak one. Thus, effect of oxygen overwhelms that of the methyl group and it is the oxygen atom that directs the incoming substituent into positions 2 and 4:

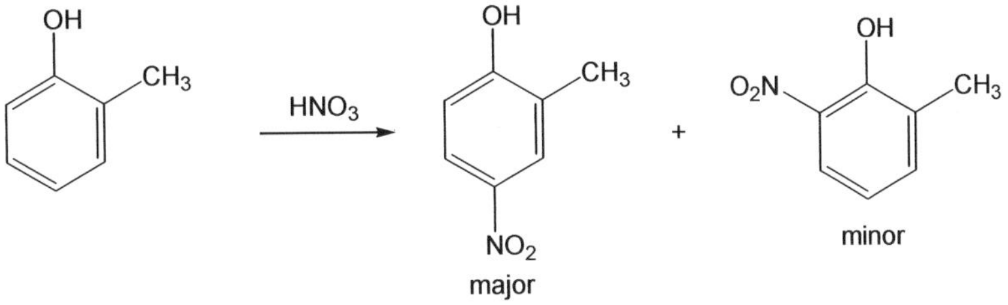

A less sterically hindered compound, in which nitro group is para to the hydroxyl substituent, will be the major product.

6. This is a bit lengthy reaction sequence and for that reason one can find this question to be rather difficult. The best way to approach it is to work it out backwards from benzonitrile to benzene. Thus, simply ignore the fact that benzene is the starting material and consider how to prepare benzonitrile. What starting material and which reaction would you use? You probably know of only one reaction to prepare benzonitrile – a reaction of a diazonium salt with copper(I) cyanide:

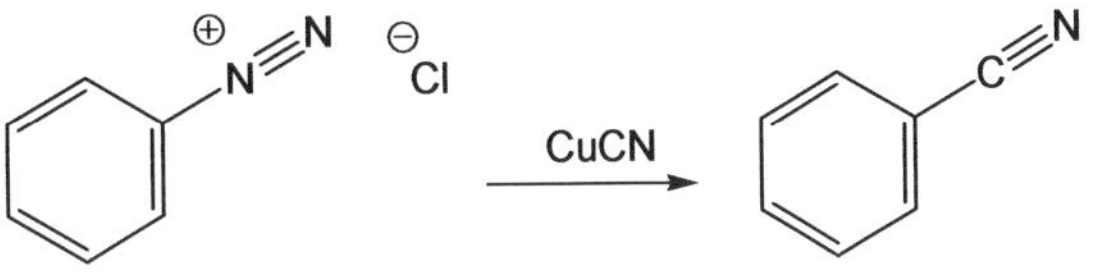

Now you have a method for preparation of the desired compound – benzonitrile, but the starting material is not "right." You have benzene as a starting material not a diazonium salt. That is OK. Now your next target is preparation of the diazonium salt. How would you prepare it?

Diazonium salts are prepared in a reaction between aniline, sodium nitrite and hydrochloric acid. Therefore:

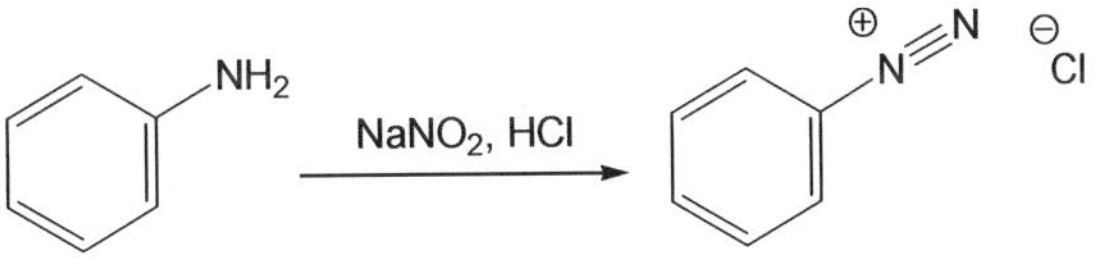

Now, your next target is preparation of aniline. How would you prepare it? You are probably aware of a couple of methods both of which involve reduction of nitrobenzene:

PhNO$_2$ →[H$_2$/Ni or 1) Zn/HCl 2) NaOH] PhNH$_2$

You are almost there! Your new target is nitrobenzene. Nitrobenzene is prepared in a nitration reaction of benzene:

C$_6$H$_6$ →[HNO$_3$, H$_2$SO$_4$] PhNO$_2$

Now you finally have benzene as a starting material. Overall reaction sequence is:

C$_6$H$_6$ →[HNO$_3$, H$_2$SO$_4$] PhNO$_2$ →[H$_2$/Ni or 1) Zn/HCl 2) NaOH] PhNH$_2$ →[NaNO$_2$, HCl] PhN$_2^+$Cl$^-$ →[CuCN] PhCN

If you are familiar with reactions of organometallic compounds and reactions of carboxylic acids and derivatives, you may have designed an alternative sequence like this:

C$_6$H$_6$ →[Br$_2$, FeBr$_3$] PhBr →[Li/hexane or Mg/ether] PhLi or PhMgBr →[1) CO$_2$ 2) HCl] PhCO$_2$H →[NH$_3$] PhCONH$_2$ →[P$_2$O$_5$] PhCN

Questions 7-9.

7. What is the product of the following reaction?

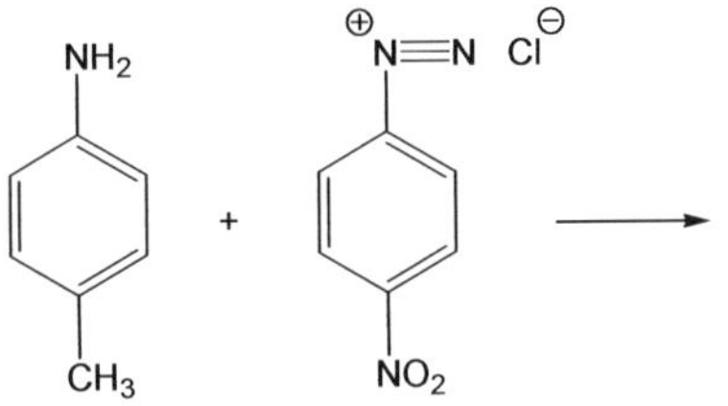

[4-methylaniline + 4-nitrobenzenediazonium chloride →]

8. Draw a structural formula of the intermediate **1** and explain its high reactivity. In what ratio are the compounds **2** and **3** formed? Explain your answer. Name the compounds **2** and **3**.

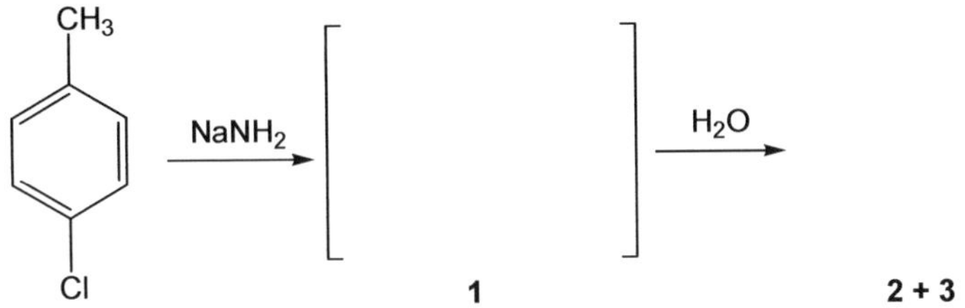

[4-chlorotoluene —NaNH$_2$→ [**1**] —H$_2$O→ **2 + 3**]

9. By what mechanism is the following intermediate formed? Explain your answer.

[chlorobenzene —NaNH$_2$→ [benzyne]]

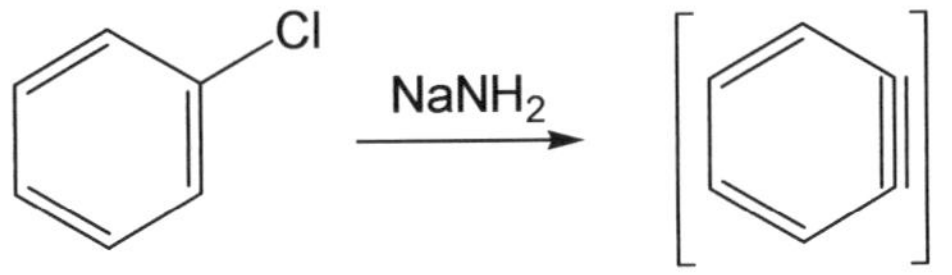

Answers to questions 7-9.

7. The reaction is a diazo coupling. Since the para position of the substituted aniline is occupied by a methyl group, the azo substituent ends up in an ortho position:

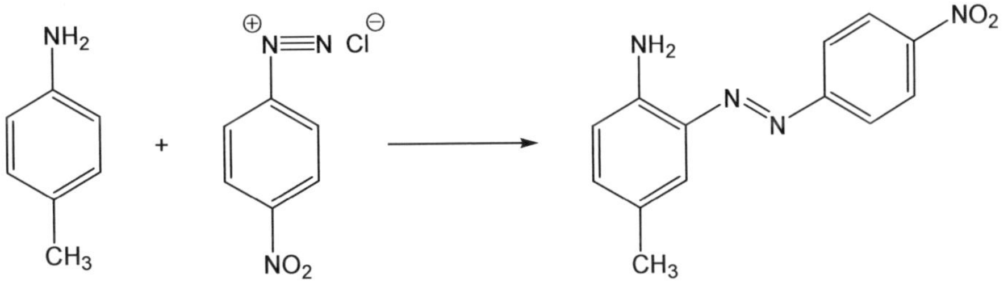

[4-methylaniline + 4-nitrobenzenediazonium chloride → 2-amino-5-methyl-4'-nitroazobenzene]

8. A reaction between a halobenzene and sodium amide gives the intermediate benzyne. Triple bond introduces a high degree of angle strain in benzyne and the intermediate is highly reactive. It reacts with nucleophiles, such as water, to give the corresponding addition products. The addition is not regioselective and the two products are obtained in approximately the same amount.

9. Benzyne is formed in an elimination reaction. The reaction mechanism is either E2 or E1cB. Both the proton and the leaving group are in a favorable, planar geometry, for the elimination. E1 mechanism is not very likely as it would involve formation of a highly unstable phenyl cation.

Questions 10-12.

10. Provide products of the following reactions:

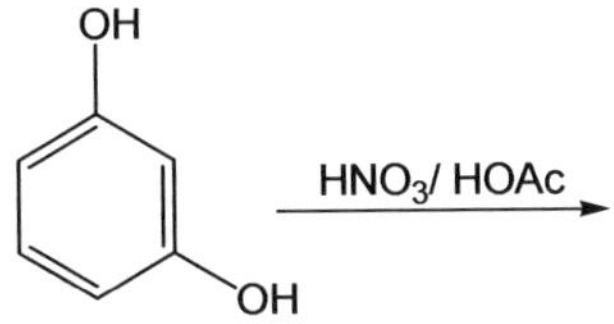

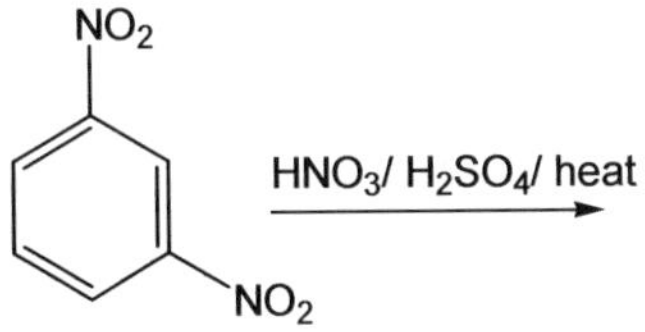

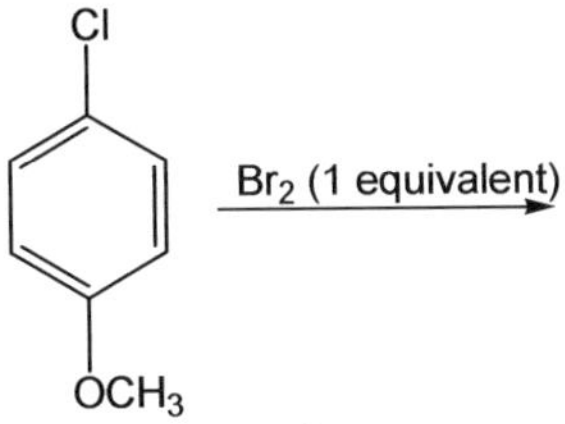

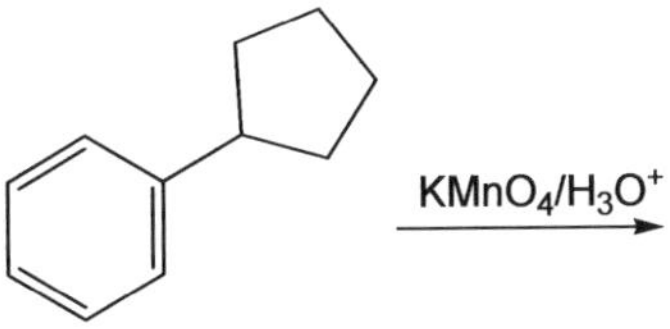

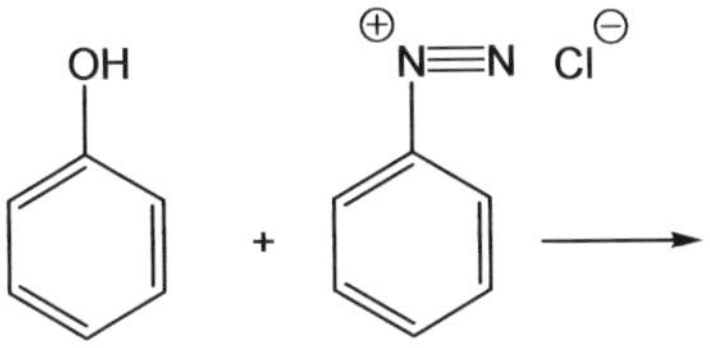

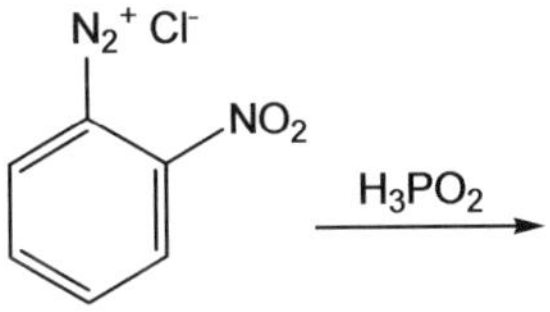

resorcinol (1,3-dihydroxybenzene) + HNO₃/ HOAc →

1,3-dinitrobenzene + HNO₃/ H₂SO₄/ heat →

4-chloro-1-methoxybenzene (with Cl up, OCH₃ down) + Br₂ (1 equivalent) →

cyclopentylbenzene + KMnO₄/H₃O⁺ →

phenol + PhN₂⁺Cl⁻ →

2-nitrobenzenediazonium chloride + H₃PO₂ →

11. Provide starting materials for the following reactions:

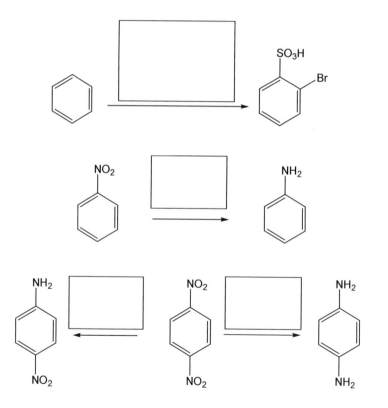

12. For the following reactions provide the reagents.

Answers to questions 10-12.

10. Reaction products:

[Resorcinol + HNO₃/HOAc → 4-nitroresorcinol]

[1,3-dinitrobenzene + HNO₃/H₂SO₄/heat → 1,3,5-trinitrobenzene]

[4-chloroanisole + Br₂ (1 equivalent) → 2-bromo-4-chloroanisole]

[cyclopentylbenzene + KMnO₄/H₃O⁺ → benzoic acid]

[phenol + benzenediazonium chloride → 4-hydroxystilbene (4-hydroxyazo coupling product shown as stilbene)]

[2-nitrobenzenediazonium chloride + H₃PO₂ → nitrobenzene]

11. Starting materials:

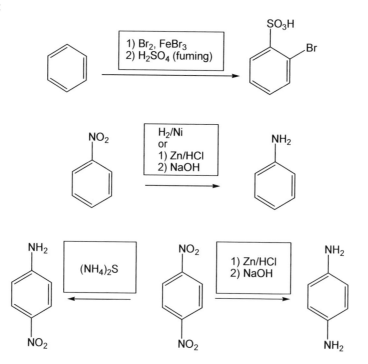

12. Reagents:

Exercises in Organic Chemistry

Additional Exercises: Questions 13-20

13. Complete the following table giving the structure or name for each molecule listed.

Name	Structural formula
propiophenone	
	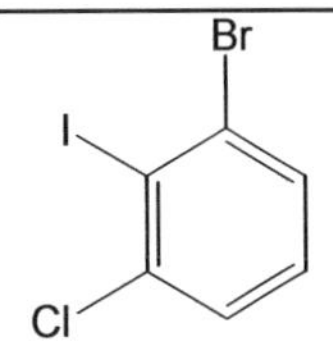 1-bromo-2-iodo-3-chlorobenzene (structure shown)
3-bromophenol	
	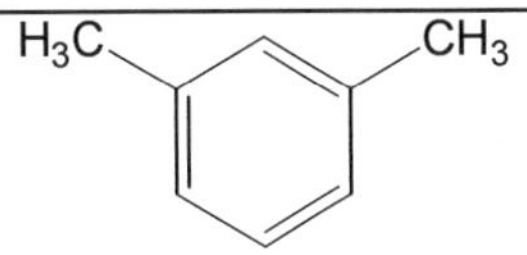 1,3-dimethylbenzene (structure shown)
2-bromo-6-fluoro-4-nitrotoluene	
	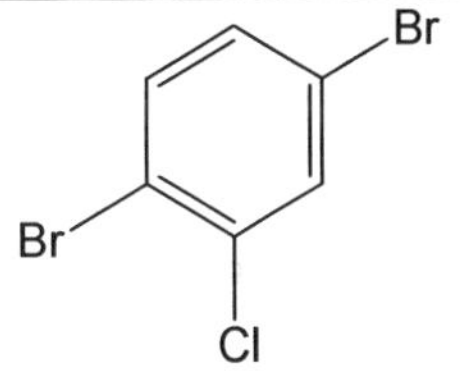 1,2-dibromo-3-chloro... (structure shown with Br, Br, Cl)

14. Provide starting materials for the following reactions.

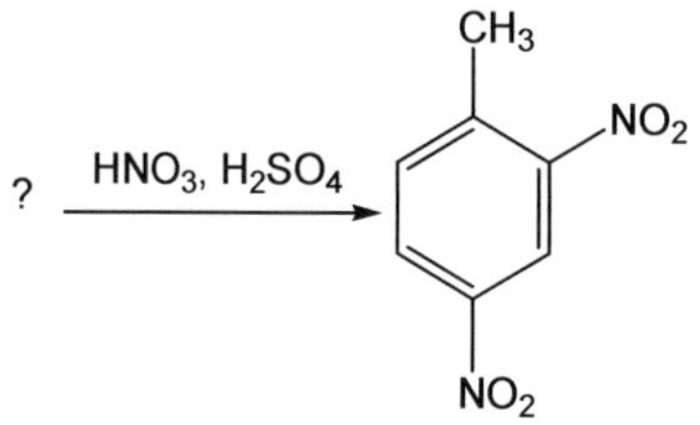

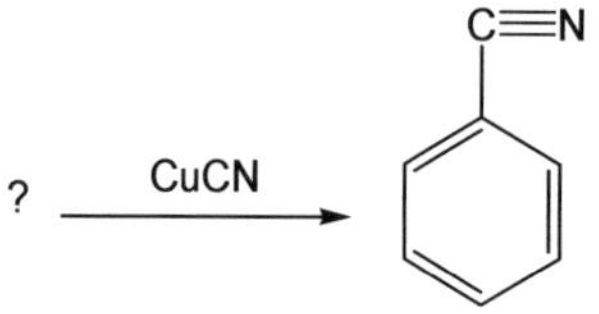

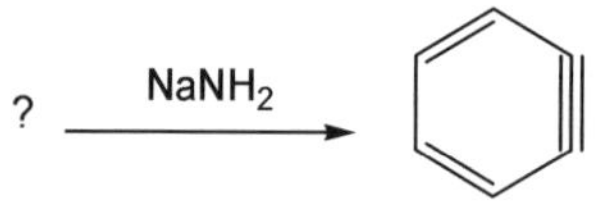

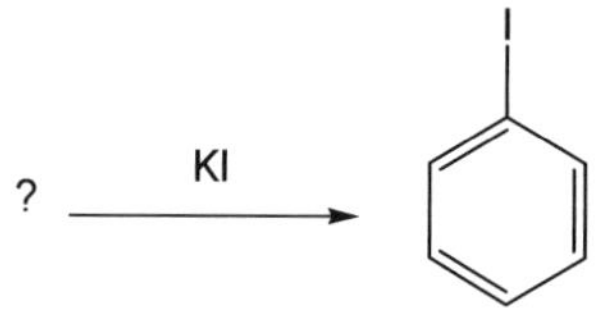

? $\xrightarrow{HNO_3, H_2SO_4}$ 2,4-dinitrotoluene (CH$_3$ with NO$_2$ at positions 2 and 4)

? $\xrightarrow{CuCN}$ benzonitrile (C≡N on benzene)

? $\xrightarrow{NaNH_2}$ benzene

? $\xrightarrow{KI}$ iodobenzene

15. Provide a structure of the missing intermediate.

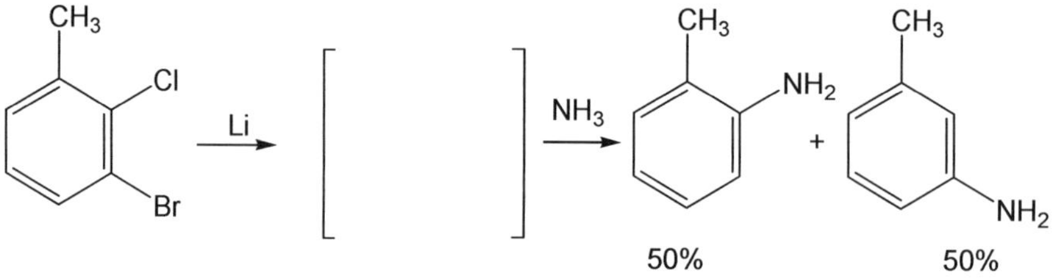

3-bromo-2-chlorotoluene $\xrightarrow{Li}$ [?] $\xrightarrow{NH_3}$ 2-methylaniline (50%) + 3-methylaniline (50%)

16. Provide reagents for the following reactions.

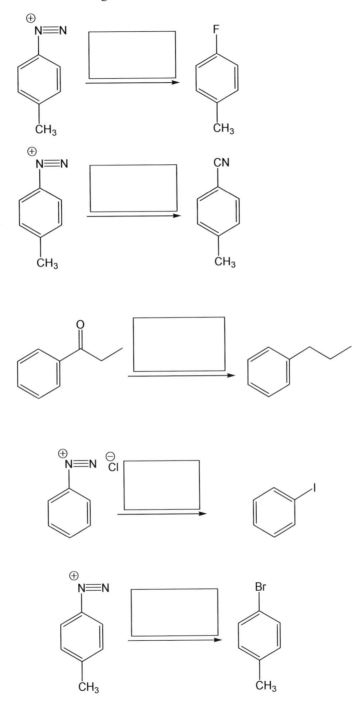

17. Provide products of the following reactions.

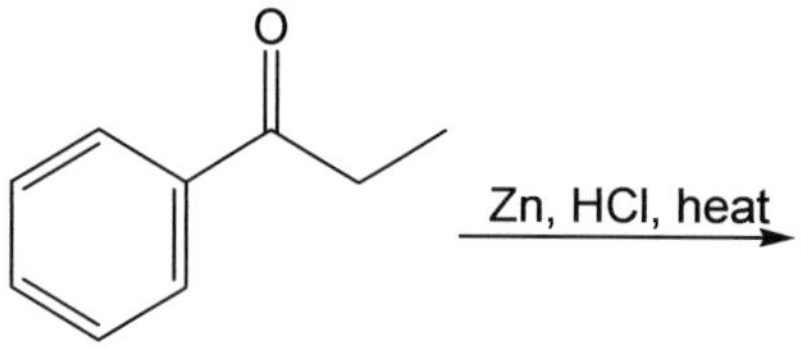

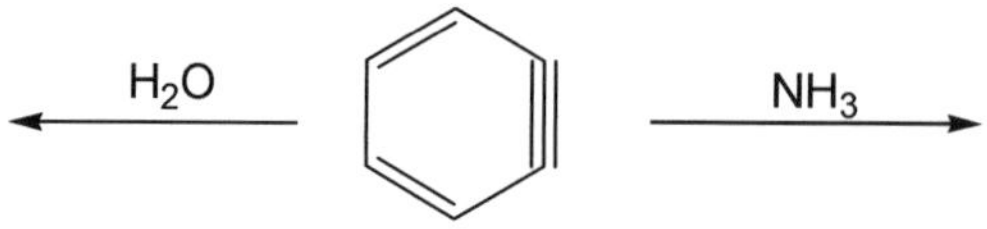

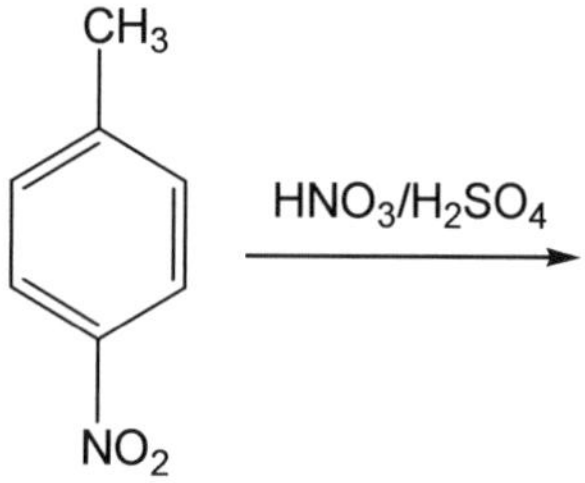

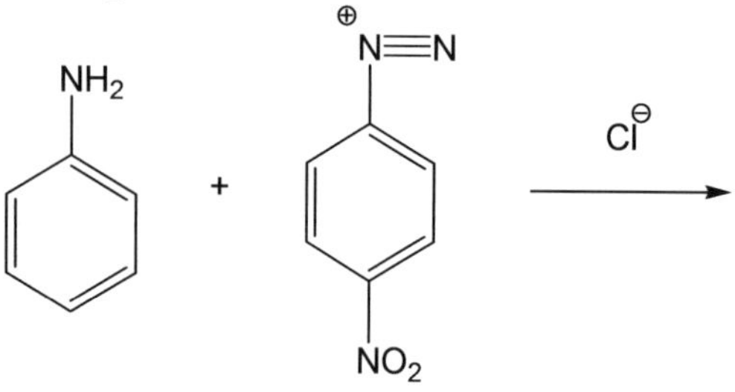

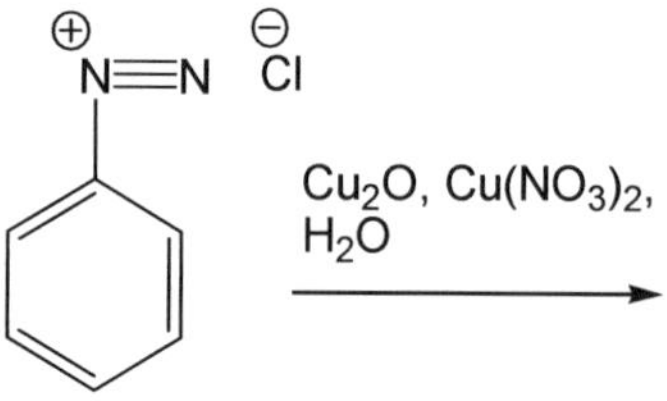

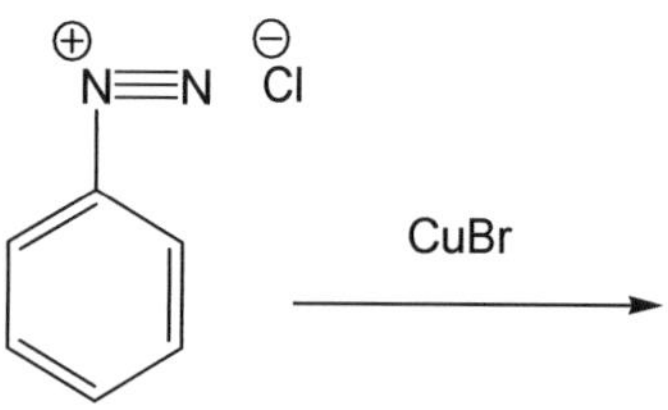

PhC(O)CH$_2$CH$_3$ $\xrightarrow{\text{Zn, HCl, heat}}$

C$_6$H$_6$ $\xleftarrow{\text{H}_2\text{O}}$ ... $\xrightarrow{\text{NH}_3}$

4-nitrotoluene $\xrightarrow{\text{HNO}_3/\text{H}_2\text{SO}_4}$

PhNH$_2$ + [4-O$_2$N-C$_6$H$_4$-N$_2$]$^+$ Cl$^-$ $\longrightarrow$

PhN$_2^+$ Cl$^-$ $\xrightarrow{\text{Cu}_2\text{O, Cu(NO}_3)_2,\ \text{H}_2\text{O}}$

PhN$_2^+$ Cl$^-$ $\xrightarrow{\text{CuBr}}$

18. Provide structures of the missing compounds:

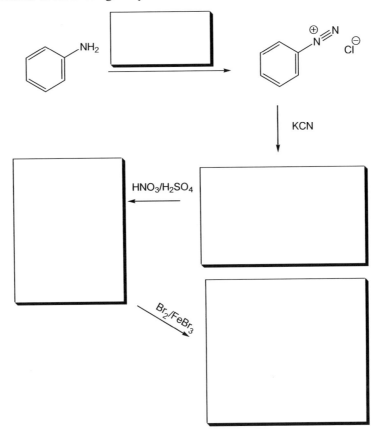

19. Which of the following gives the correct order of strength of the following acids from the weakest to the strongest?

I: 4-nitro-2-... (O$_2$N, NO$_2$, OH on benzene)
II: 2-nitrophenol (NO$_2$, OH)
III: 2-nitro-5-methylphenol (NO$_2$, H$_3$C, OH)
IV: 2-methylphenol (CH$_3$, OH)

a) I<II<IV<III
b) II<I<IV<III
c) III<IV<I<II
d) III<IV<II<I

20. What is the final product of the following sequence of reactions?

benzene → 1) Br₂, FeBr₃; 2) HNO₃, H₂SO₄; 3) H₂SO₄ (fuming) → ?

a) 4-bromo-3-nitro-benzenesulfonic acid (HO₃S, Br, NO₂)
b) (O₂N, Br, SO₃H)
c) (HO₃S, NO₂, Br)
d) (O₂N, SO₃H, Br)

Answers to Additional Exercises: Questions 13-20

13. Table:

Name	Structural formula
propiophenone	C₆H₅–C(=O)–CH₂CH₃
1-bromo-3-chloro-2-iodobenzene	benzene with Br, I, Cl at 1,2,3
3-bromophenol	phenol with Br at 3-position
m-xylene	1,3-dimethylbenzene
2-bromo-6-fluoro-4-nitrotoluene	toluene with F (6), Br (2), NO₂ (4)

| 1,4-dibromo-2-chlorobenzene | (structure: benzene with Br at 1, Cl at 2, Br at 4) |

14. Starting materials:

(toluene) + HNO₃, H₂SO₄ → 2,4-dinitrotoluene

PhN₂⁺Cl⁻ + CuCN → benzonitrile (PhCN)

bromobenzene + NaNH₂ → benzyne

PhN₂⁺Cl⁻ + KI → iodobenzene

15. Structure of the intermediate:

3-bromo-2-chlorotoluene + Li → [3-methylbenzyne] + NH₃ → 2-methylaniline (50%) + 3-methylaniline (50%)

16. Reagents:

- 4-methylbenzenediazonium + HBF₄, heat → 4-fluorotoluene
- 4-methylbenzenediazonium + CuCN → 4-methylbenzonitrile
- PhCOCH₂CH₃ + Zn/HCl, heat or H₂NNH₂, KOH, heat → PhCH₂CH₂CH₃
- benzenediazonium chloride + KI → iodobenzene
- 4-methylbenzenediazonium + CuBr → 4-bromotoluene

17. Reaction products:

PhCOCH$_2$CH$_3$ $\xrightarrow{\text{Zn, HCl, heat}}$ PhCH$_2$CH$_2$CH$_3$

PhOH $\xleftarrow{\text{H}_2\text{O}}$ C$_6$H$_6$ $\xrightarrow{\text{NH}_3}$ PhNH$_2$

4-nitrotoluene $\xrightarrow{\text{HNO}_3/\text{H}_2\text{SO}_4}$ toluene

PhNH$_2$ + 4-O$_2$N-C$_6$H$_4$-N$_2^+$ Cl$^-$ $\longrightarrow$ 4-H$_2$N-C$_6$H$_4$-N=N-C$_6$H$_4$-NO$_2$-4

PhN$_2^+$ Cl$^-$ $\xrightarrow{\text{Cu}_2\text{O, Cu(NO}_3)_2,\ \text{H}_2\text{O}}$ PhOH

PhN$_2^+$ Cl$^-$ $\xrightarrow{\text{CuBr}}$ PhBr

18. Missing compounds:

19. d); 20. b).

Exercises in Organic Chemistry

Test 5: CHAPTERS 16-19

1. Arrange the following free radicals in the order of increasing stability.

2. Provide resonance structures for phenoxide ion.

3. Predict which of the two alkenes, shown below, is going to be more reactive towards HBr. Explain your answer.

4. Which of the following groups has an electron-withdrawing resonance effect?

   a) $-CH_2OCH_3$
   b) $-F$
   c) $-OCH_3$
   d) $-NO_2$

5. Provide a drawing of molecular orbitals of benzene. Indicate the relative energy levels.

6. Draw structure of the major organic product (if any) in each case:

7. Complete the following table giving the structure or a *complete* name for each molecule listed.

	(structure shown)
(Z)-2-methyl-1,3-pentadiene	
	(structure shown)
(S)-3-bromo-1-pentanol	
	(structure shown)

8. Apply Hückel's rule and predict which of the following hydrocarbons will exhibit aromatic stabilization. Show your work.

9. Draw resonance structures of naphthalene.

10. Provide resonance structures for cyclopentadienyl anion.

11. Provide a mechanism for halogenation of benzene.

12. How would you prepare propylbenzene starting with benzene and any other organic or inorganic reagents?

13. Complete the following table giving the structure or name for each molecule listed.

Name	Structural formula
	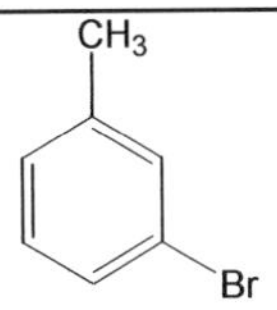 3-bromotoluene (CH₃ and Br in meta positions)
2,6-dinitrophenol	
	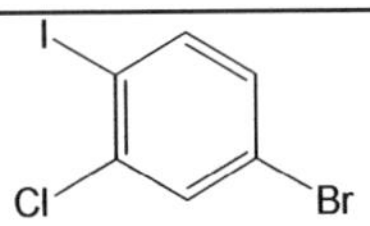 1-bromo-3-chloro-4-iodobenzene (structure with I, Cl, Br)
m-xylene	
	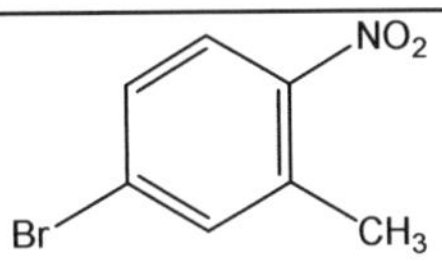 4-bromo-2-methyl-1-nitrobenzene (Br, CH₃, NO₂ substituents)
2,3,5-trichloroanisole	

14. Draw the structure of the major organic product (if any) in each case:

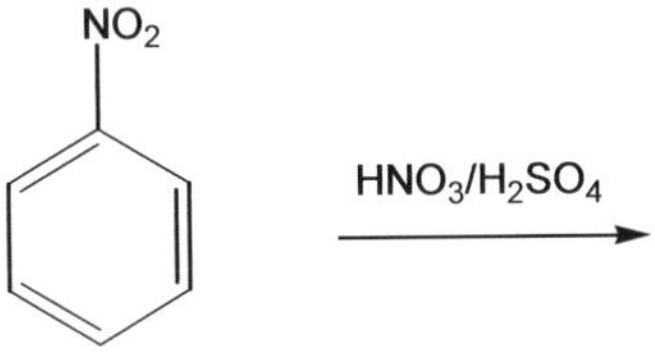

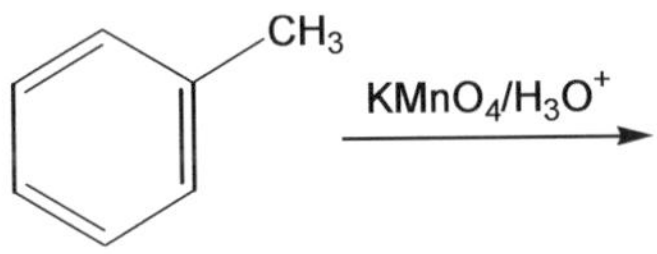

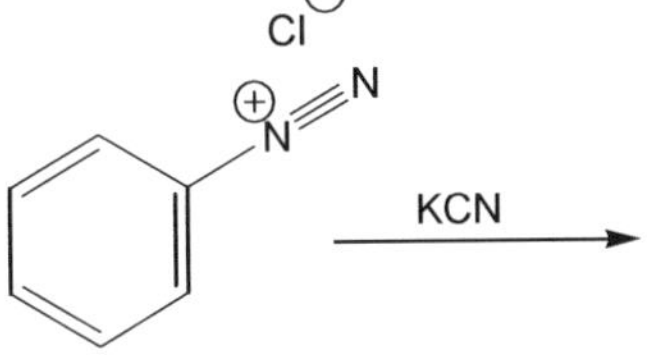

nitrobenzene + HNO₃/H₂SO₄ →

toluene + KMnO₄/H₃O⁺ →

benzenediazonium chloride (PhN₂⁺ Cl⁻) + KCN →

219

15. How would you prepare the following compounds:

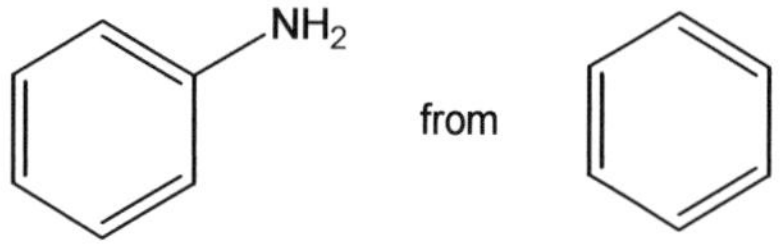

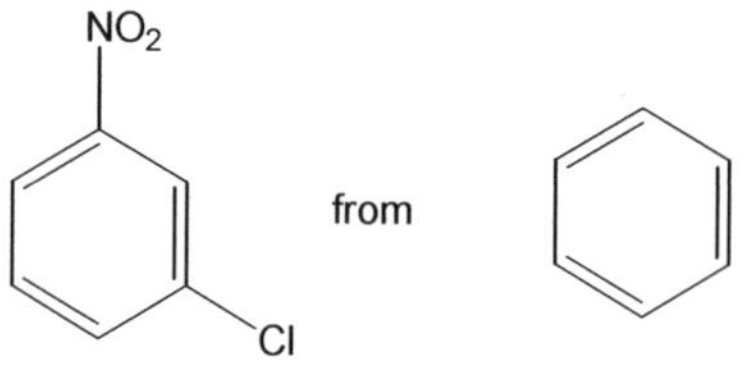

   PhNH$_2$ from benzene

   3-chloronitrobenzene from benzene

16. Explain why is chloride substituent deactivating and ortho, para- directing?

17. Draw resonance structures of nitrobenzene.

18. Provide the structure of missing intermediate. Explain why are *ortho-* and *meta*-aminoanisole obtained in equal amounts?

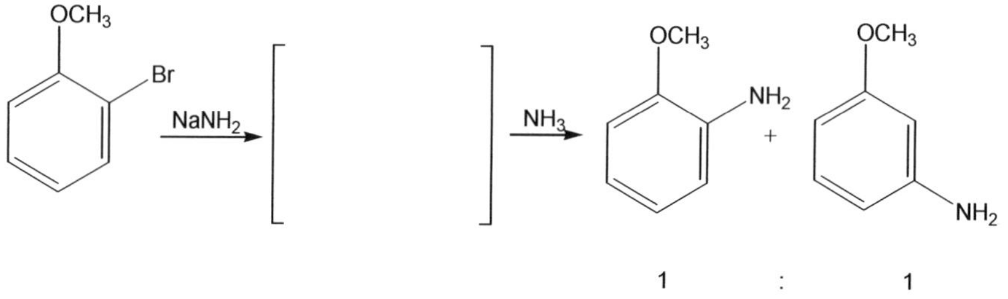

   2-bromoanisole $\xrightarrow{NaNH_2}$ [ ] $\xrightarrow{NH_3}$ *ortho*-aminoanisole + *meta*-aminoanisole

   1 : 1

19. Draw the structures of missing compounds:

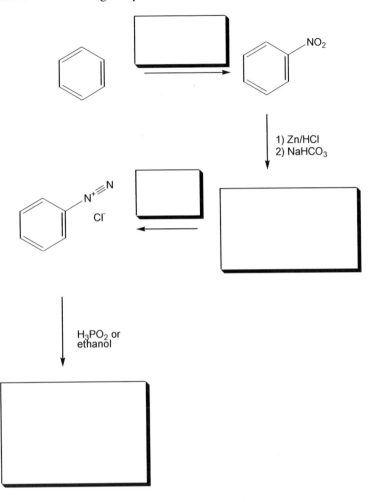

20. Classify the following substituents as activating or deactivating and state whether activity of each substituent is strong, moderate or weak.

—Br   —C(=O)—OCH$_3$ —SO$_3$H   —CH$_2$CH$_3$ —O—C(=O)—CH$_3$

Test 5 Solutions

1. Arrange the following free radicals in the order of increasing stability.

   n-propyl • < tert-butyl • < benzyl •

2. Provide resonance structures for phenoxide ion.

3. Reactivity towards HBr:

   3,4-dihydro-2H-pyran (+R, −I) more reactive

   3,6-dihydro-2H-pyran (−I only)

4. d)

5. Molecular orbitals of benzene:

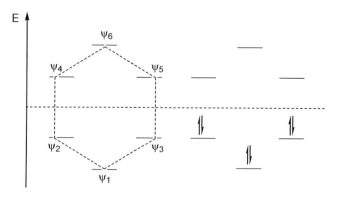

Exercises in Organic Chemistry

6. Draw the structure of the major organic product (if any) in each case:

7. Completed table.

Name	Structure
(E)-2-hepten-5-yne	
(Z)-2-methyl-1,3-pentadiene	
(2Z,5E)-3-methyl-2,5-octadiene	
(S)-3-bromo-1-pentanol	
(E)-2-methyl-1,3-pentadiene	

8. Aromatic and non-aromatic hydrocarbons:

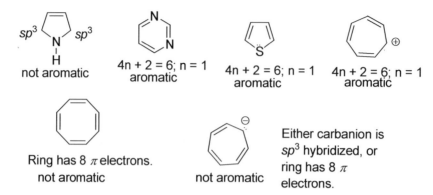

9. Resonance structures of naphthalene:

10. Provide resonance structures for cyclopentadienyl anion.

11. Provide a mechanism for halogenation of benzene.

Exercises in Organic Chemistry

12. How would you prepare propylbenzene starting with benzene and any other organic or inorganic reagents?

[benzene] + [propanoyl chloride] →(1) AlCl₃, 2) H₂O)→ [propiophenone] →(Zn/Hg, HCl, heat or H₂NNH₂, KOH, heat)→ [propylbenzene]

13. Completed table:

Name	Structural formula
m-bromotoluene, or 3-bromotoluene	[benzene ring with CH₃ and Br in meta positions]
2,6-dinitrophenol	[phenol with NO₂ groups at 2,6 positions]
4-bromo-2-chloro-1-iodobenzene	[benzene with I, Cl, Br at 1,2,4 positions]
m-xylene	[benzene with two CH₃ in meta positions]
5-bromo-2-nitrotoluene	[benzene with Br, NO₂, CH₃]
2,3,5-trichloroanisole	[benzene with OCH₃ and three Cl]

14. Reaction products:

[Reaction 1: nitrobenzene + HNO₃/H₂SO₄ → 1,3-dinitrobenzene]

[Reaction 2: toluene + KMnO₄/H₃O⁺ → benzoic acid]

[Reaction 3: benzenediazonium chloride + KCN → benzonitrile]

15. Preparation:

[Benzene → (HNO₃/H₂SO₄) → nitrobenzene → 1) Zn/HCl, 2) NaOH → aniline]

[Benzene → 1) HNO₃, H₂SO₄; 2) Cl₂, AlCl₃ → 1-chloro-3-nitrobenzene]

16. Chloride is deactivating because it is an electron withdrawing substituent, due to high electronegativity of chlorine atom. It is ortho, para directing because it stabilizes positive charge on the phenyl ring when the incoming substituent adds to either ortho or para position. It does not stabilize the intermediate cyclohexadienyl cation when the incoming substituent is in the meta position.

17. Resonance structures of nitrobenzene:

[Five resonance structures of nitrobenzene shown with curved arrows indicating electron movement between structures]

226

18. The structure of missing intermediate:

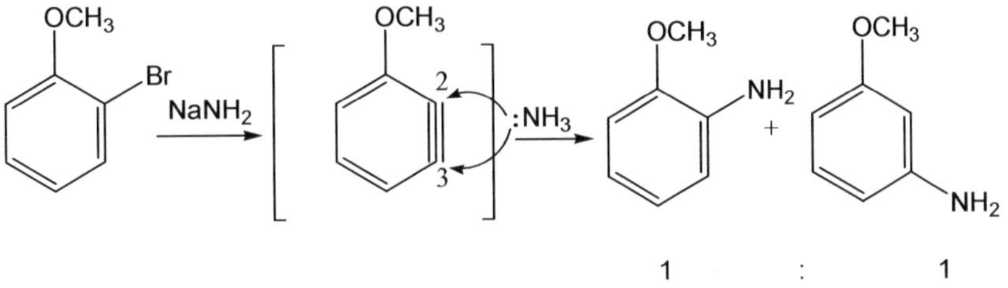
OCH3
Br
NaNH2
OCH3
2
3
:NH3
OCH3
NH2
+
OCH3
NH2
1
:
1

Attack of ammonia (NH_3) nucleophile occurs at the positions 2 and 3 of the intermediate benzyne with equal probability.

19. The structures of missing compounds:

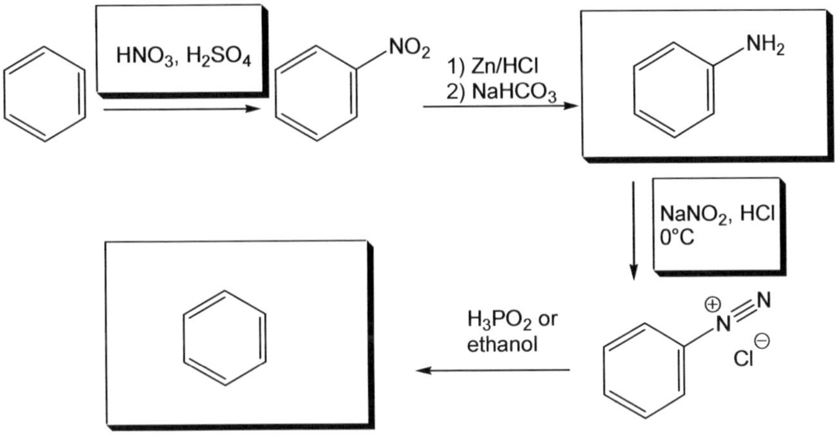
HNO3, H2SO4
NO2
1) Zn/HCl
2) NaHCO3
NH2
NaNO2, HCl
0°C
N
N
Cl
H3PO2 or
ethanol

20. Classification of the substituents:

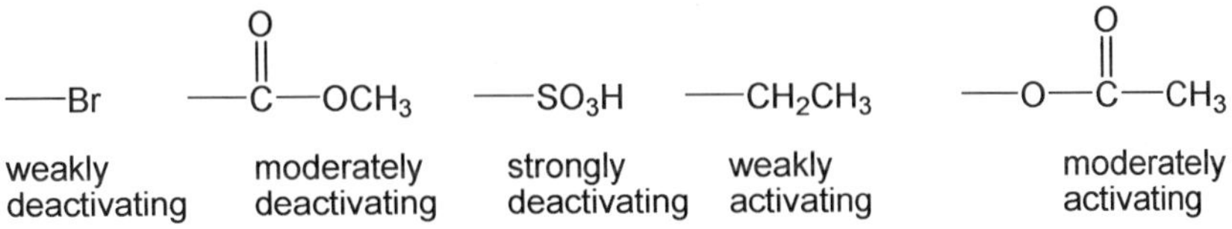
—Br
weakly
deactivating
O
—C—OCH3
moderately
deactivating
—SO3H
strongly
deactivating
—CH2CH3
weakly
activating
O
—O—C—CH3
moderately
activating

Exercises in Organic Chemistry

20. Carbonyl Compounds I: Aldehydes and Ketones

Concepts to review: nomenclature of aldehydes and ketones, reactivity of carbonyl compounds, carbon nucleophiles, organometallic reagents, acetylide and cyanide ions, Wittig reaction, hydride ion, oxygen nucleophiles, nitrogen nucleophiles, addition to α,β-unsaturated carbonyl compounds.

Questions 1-5.

1. Name each of the following aldehydes and ketones according to both systematic (IUPAC) as well as common nomenclature. Note that for some compounds you may not be able to come up with a common name. You should be able to name all of them according to systematic nomenclature.

2. Arrange the following compounds in the order of increasing reactivity.

3. What is the product of a reaction of ethylmagnesium bromide and 2-pentanone?

4. What is the product of reaction of butyllithium and carbon dioxide?

5. What reagents would you use for the synthesis of 3-ethyl-1-pentene by means of a Wittig reaction?

Answers to questions 1-5.

1. **IUPAC and common names:**

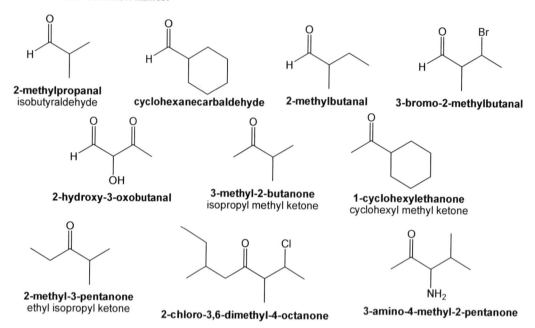

2. Reactivity of aldehydes and ketones depends mainly on the degree of substitution next to the carbonyl carbon. Do not be confused by the rings! It is still a degree of substitution that you should consider. Least substituted, and least hindered, compound is the most reactive, while the most substituted is the least reactive. Therefore, compound I with two tertiary carbons next to carbonyl is the least reactive, while aldehyde II, with only one secondary carbon as a substituent, is the most reactive. Overall order of reactivity, from the least reactive to the most reactive is: I < III < IV < II.

3. Common reaction of aldehydes and ketones is a nucleophilic addition. To answer a question like this, one has to identify the nucleophile. The nucleophile that makes a bond with the carbonyl carbon and the oxygen is protonated to give the corresponding hydroxyl group. In this particular reaction carbon of ethylmagnesium bromide is the nucleophile and forms a bond with the carbonyl carbon of 2-pentanone. The final product is a tertiary alcohol.

4. In this case carbon of butyllithium bonded to the lithium is the nucleophile and carbon dioxide is the carbonyl compound. Carbon dioxide has two carbonyl groups and only one of the two reacts with a nucleophile. Upon protonation, the final product is the corresponding carboxylic acid. Note that the carboxylic acid has one more carbon atom than the starting organometallic compound (the nucleophile).

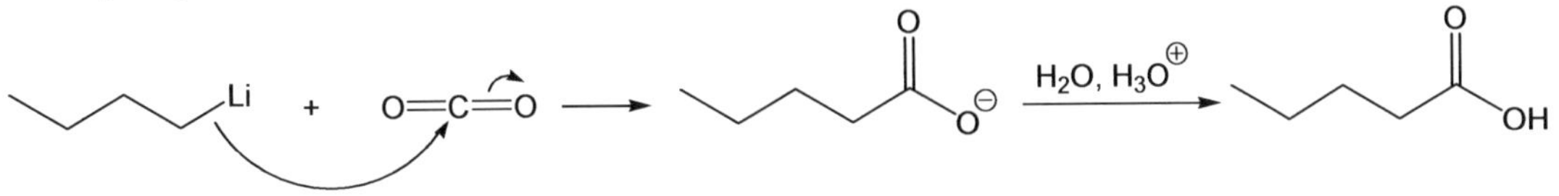

5. One should start with a structural formula of 3-ethyl-1-pentene:

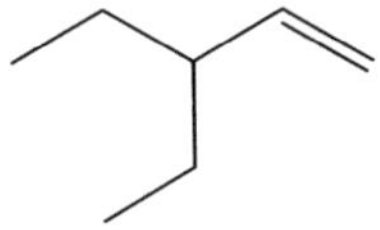

In the course of a Wittig reaction a carbonyl compound a a phosphorus ylide react to form a new carbon-carbon double bond. Therefore, there are two sets of reactants that can give the desired product:

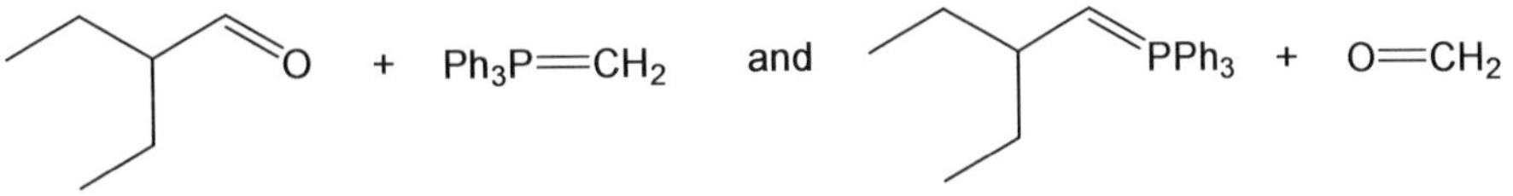

$$\text{structure} + Ph_3P=CH_2 \quad \text{and} \quad \text{structure} = PPh_3 + O=CH_2$$

Questions 6-9.

6. Provide starting materials for the following reactions.

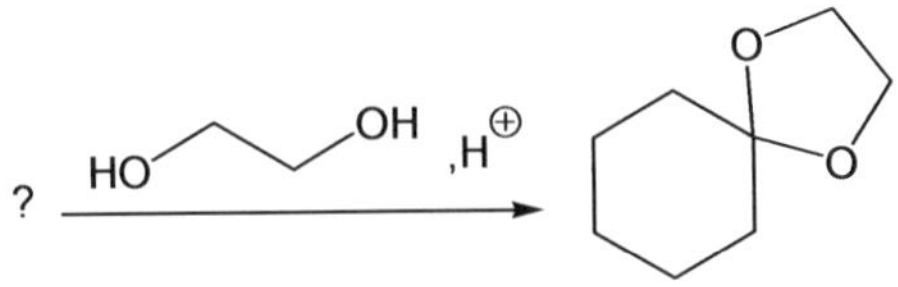

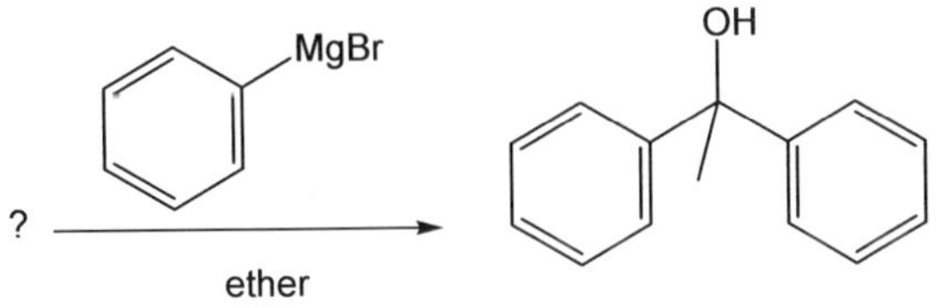

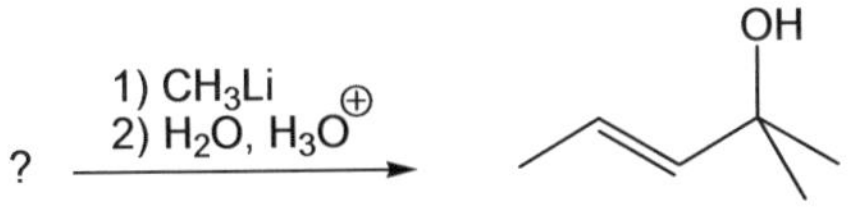

7. Provide products of the following reactions.

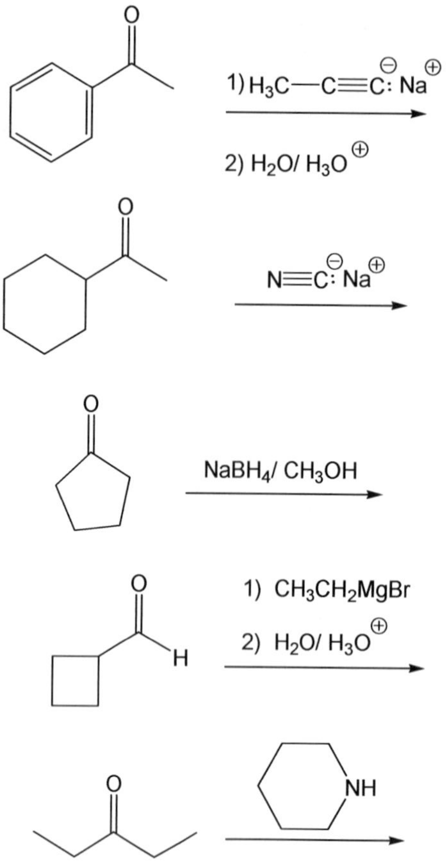

PhC(O)CH₃
1) H₃C—C≡C:⁻ Na⁺
2) H₂O / H₃O⁺

Cyclohexyl-C(O)CH₃
N≡C:⁻ Na⁺

Cyclopentanone
NaBH₄ / CH₃OH

Cyclobutyl-CHO
1) CH₃CH₂MgBr
2) H₂O / H₃O⁺

Pentan-2-one (CH₃CH₂C(O)CH₂CH₃) + piperidine (NH)

Exercises in Organic Chemistry

8. Provide reagents for the following reactions.

(isoamyl-MgBr) → [ ] → (isohexanoic acid, CH₂CH₂CO₂H derivative)

(pentan-3-one) → [ ] → (cyclic ketal, 1,3-dioxane)

(cyclopentanone) → [ ] → (cyclopentanone phenylhydrazone, =N-NH-Ph)

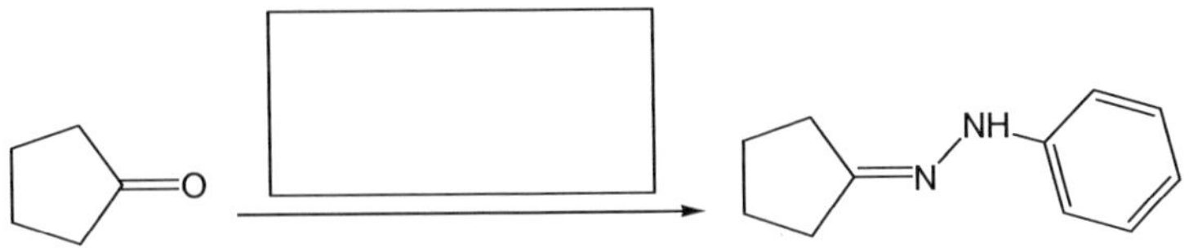

9. What are the products of the following reactions?

(cyclobutyl ethyl ketone) + NH_2OH →

(cyclobutyl ethyl ketone) + pyrrolidine →

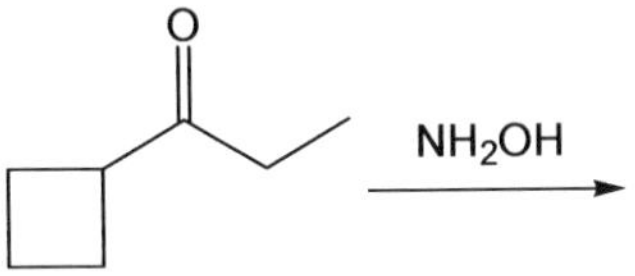

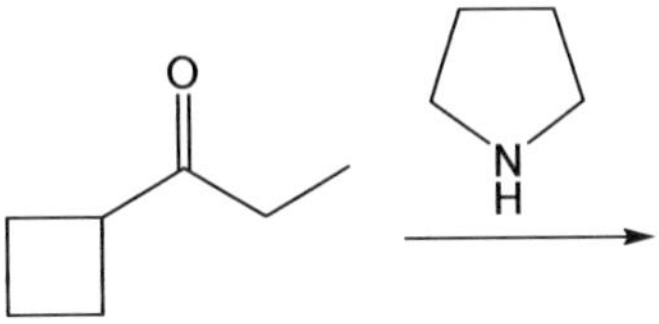

Answers to questions 6-9.

6. Starting materials:

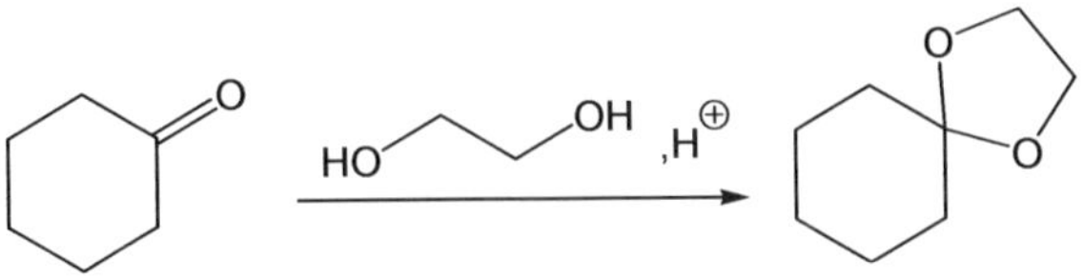

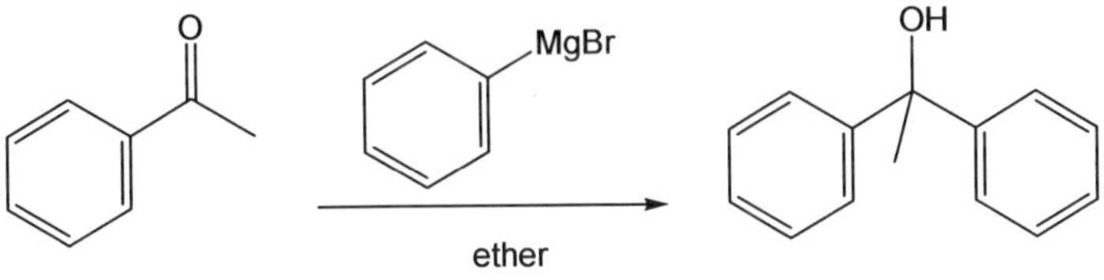

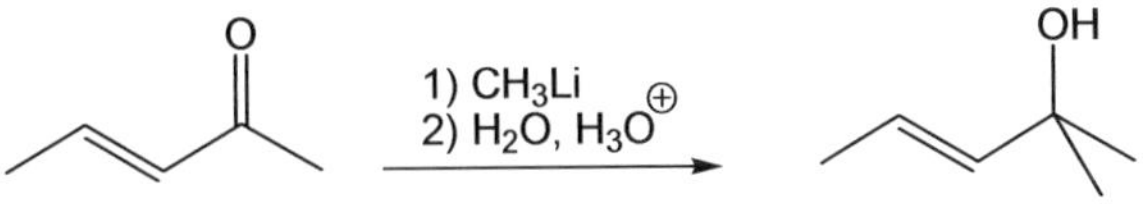

cyclohexanone + HO-CH₂CH₂-OH, H^+ → cyclohexanone ethylene ketal

acetophenone + PhMgBr, ether → 1,1-diphenylethanol (Ph₂C(OH)CH₃)

(but-3-en-2-one / methyl vinyl ketone type) 1) CH_3Li 2) H_2O, H_3O^+ → tertiary allylic alcohol

233

7. Reaction products:

8. Reagents:

9. Nitrogen nucleophiles react with a carbonyl group to give a carbon-nitrogen double bond, provided that they are capable of forming a carbon-nitrogen double bond. Only –NH_2 nucleophiles are as they have two hydrogens that can be removed in the course of a reaction and be replaced by a double bond. Nitrogen nucleophiles with only one hydrogen substituent, -NH-, give enamines, while tertiary amines do not react.

Veljko Dragojlovic

Additional Exercises: Questions 10-15

10. Complete the following table giving the structure or name for each molecule listed.

Name	Structural formula
3,3-dimethoxyhexane	
	Cl–C(CH₃)₂–C≡N
3-oxobutanal	
	CH₂=CH–C(=O)–CH₂CH₃
1,1-diethoxycyclopentane	
	1,1-diethoxycyclopropane structure (cyclopropane with two OEt groups on C1)
2,4-dimethylhexanedial	
	CH₃CH₂–C(=O)–C(H)=C(H)–CH₃

11. Provide formulas of the missing compounds.

[cyclopentanone with methyl substituent] ← ☐ → [cyclopentenone] → ☐ → [HO-cyclopentadiene with methyl]

☐ —CH₃OH (excess), H⁺→ [cyclohexane with C(OCH₃)₂ group: H₃CO, OCH₃]

12. Provide formulas of products of the following reactions.

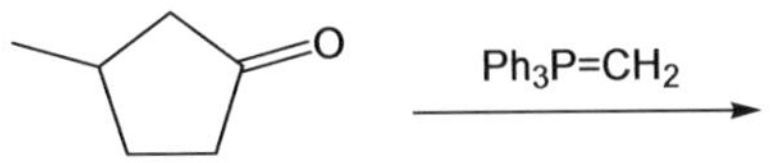

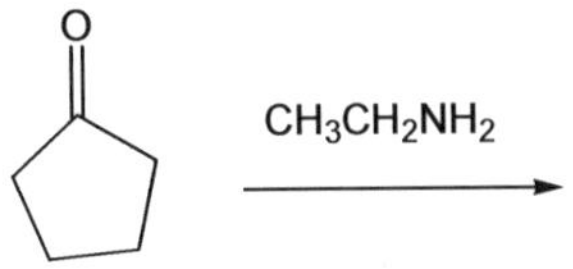

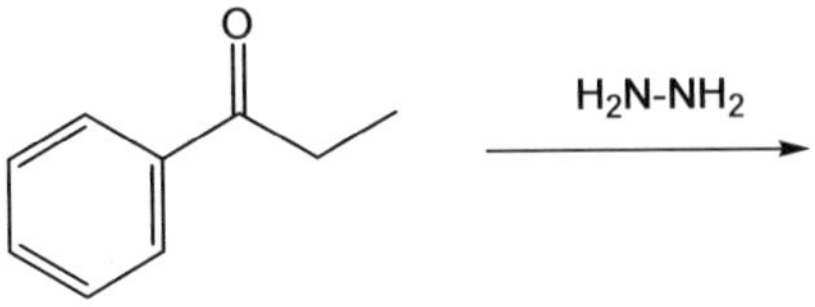

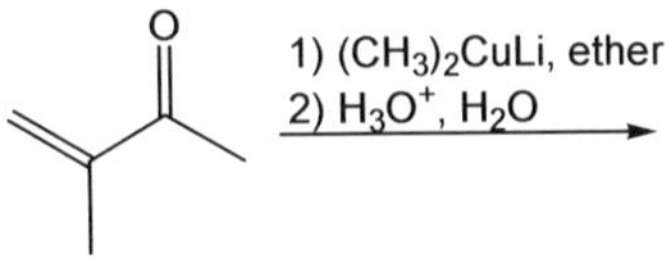

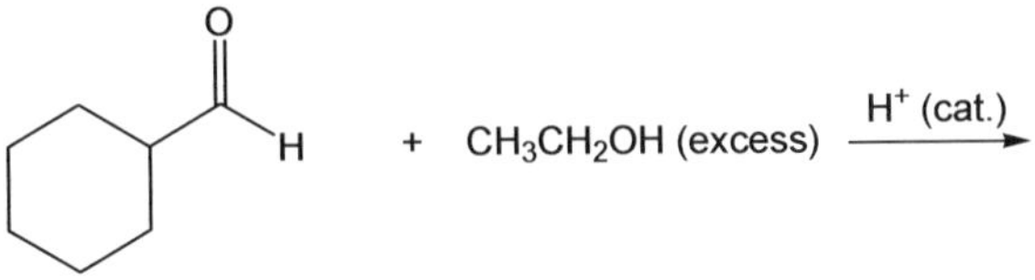

3-methylcyclopentanone + Ph₃P=CH₂ →

cyclopentanone + CH₃CH₂NH₂ →

1-phenyl-1-propanone (propiophenone) + H₂N-NH₂ →

methacryloyl methyl ketone (3-methyl-3-buten-2-one derivative: CH₂=C(CH₃)–C(=O)–CH₃... shown as methyl vinyl ketone with α-methyl)
1) (CH₃)₂CuLi, ether
2) H₃O⁺, H₂O →

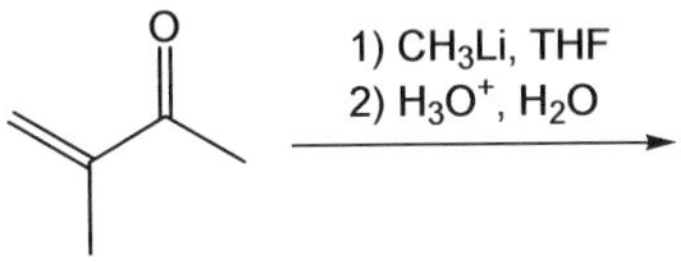

same α,β-unsaturated ketone
1) CH₃Li, THF
2) H₃O⁺, H₂O →

cyclohexanecarbaldehyde + CH₃CH₂OH (excess) —H⁺ (cat.)→

13. A source of hydride ions for reduction of carbonyl compounds is:

a) NaBH₄
b) LiH
c) CaH₂
d) H₂O

14. Cyanide ion adds to an α,β–unsaturated carbonyl compound to give:

a) 1,2-addition product
b) 1,4-addition product
c) A mixture of 1,2- and 1,4-addition products
d) An enamine

15. A ketone reacts with hydrazine to give:

   a) A Schiff base
   b) An oxime
   c) A hydrazone
   d) An imine

Additional Exercises: Questions 10-15

10. Table:

Name	Structural formula
3,3-dimethoxyhexane	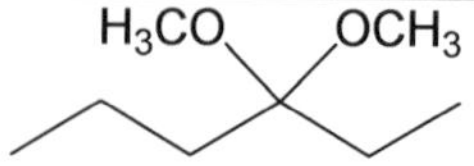
2-chloropropanenitrile	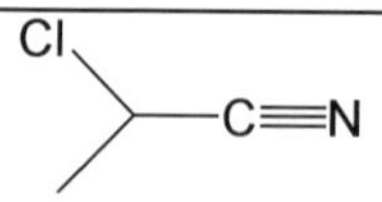
3-oxobutanal	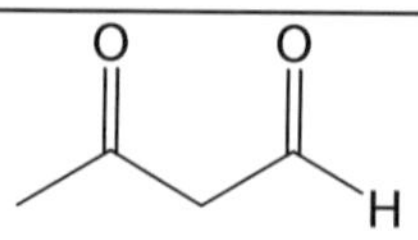
1-penten-3-one	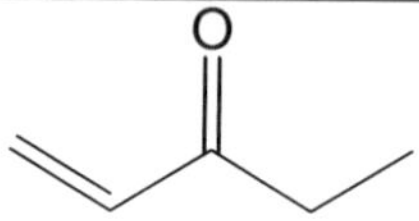
1,1-diethoxycyclopentane	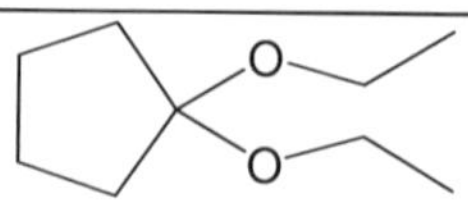
1,1-diethoxycyclopropane	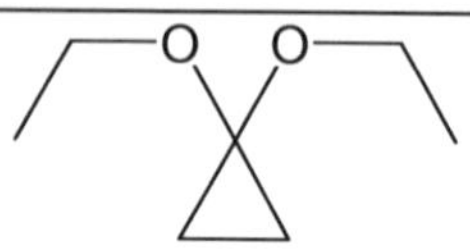
2,4-dimethylhexanedial	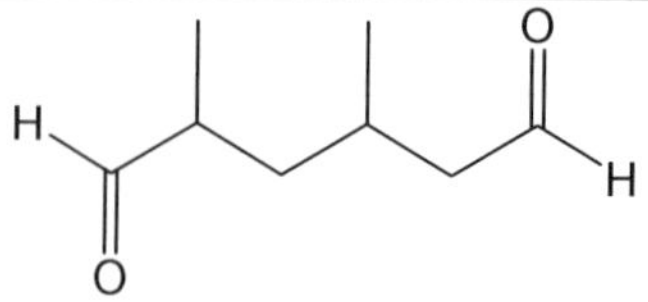
(Z)-4-hexen-3-one	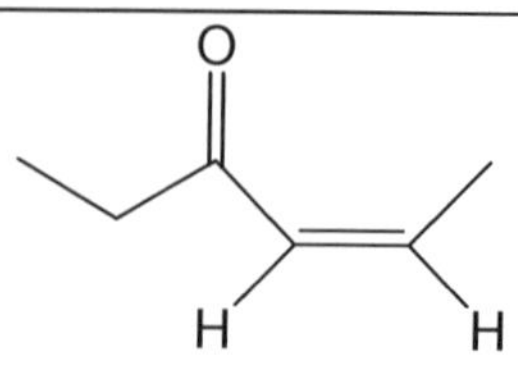

11. Missing compounds:

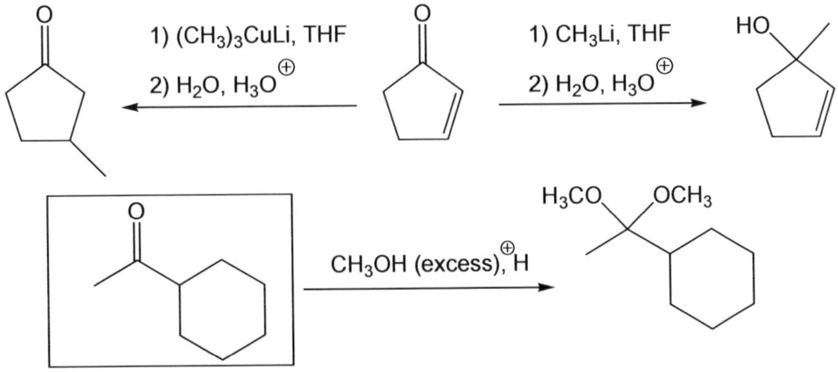

[Reaction scheme: 3-methylcyclopentanone formed from cyclopentenone via 1) (CH₃)₂CuLi, THF; 2) H₂O, H₃O⁺. Cyclopentenone reacts with 1) CH₃Li, THF; 2) H₂O, H₃O⁺ to give 1-methyl-cyclopent-2-enol (HO).]

Missing compound (boxed): 1-cyclohexyl-ethanone (acetylcyclohexane) → CH₃OH (excess), H⁺ → dimethyl ketal (H₃CO, OCH₃ on cyclohexyl).

12. Reaction products:

3-methylcyclopentanone + Ph₃P=CH₂ → 3-methyl-methylenecyclopentane

cyclopentanone + CH₃CH₂NH₂ → N-ethyl cyclopentanimine

PhC(O)CH₂CH₃ + H₂N-NH₂ → PhC(=N-NH₂)CH₂CH₃ (hydrazone)

methyl vinyl ketone analog (CH₂=C(CH₃)-C(O)-CH₃)
1) (CH₃)₂CuLi, ether
2) H₃O⁺, H₂O
→ (CH₃)₃C-C(O)-CH₃

Same enone
1) CH₃Li, THF
2) H₃O⁺, H₂O
→ CH₂=C(CH₃)-C(OH)(CH₃)₂

cyclohexanecarbaldehyde + CH₃CH₂OH (excess) → H⁺ (cat.) → cyclohexyl-CH(OEt)₂ (diethyl acetal)

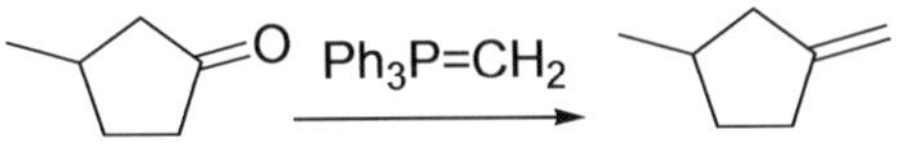

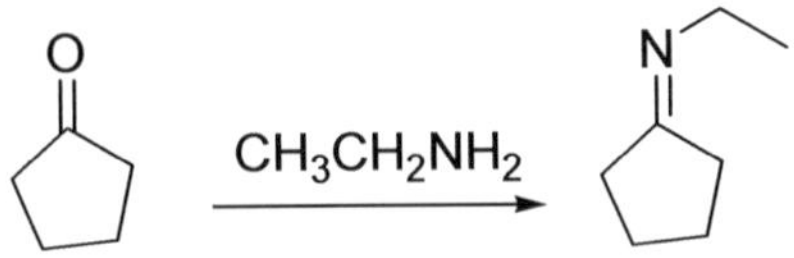

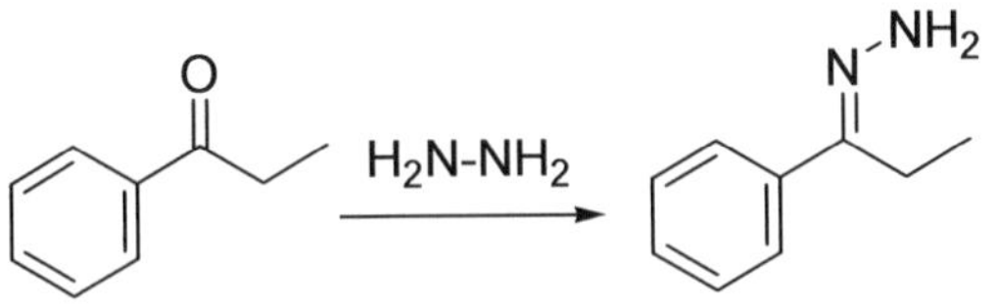

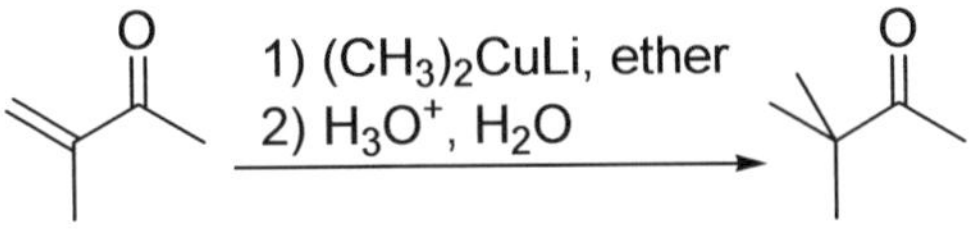

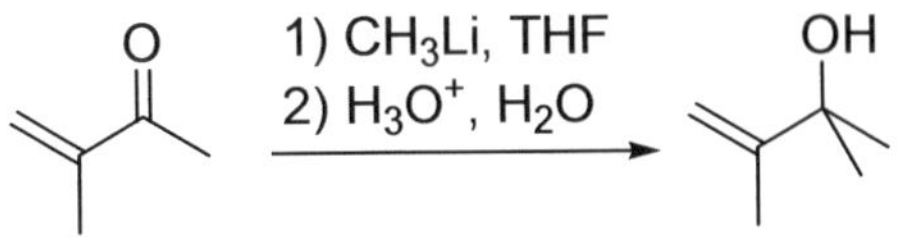

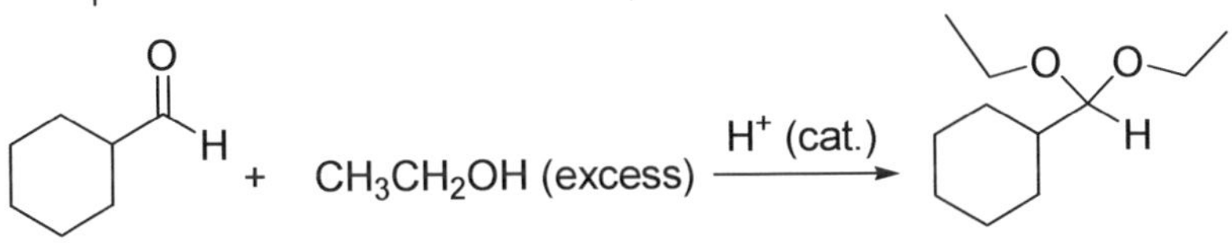

13. a); 14. b); 15. c).

21. Carbonyl Compounds II: Carboxylic Acids and Derivatives

Concepts to review: nomenclature of carboxylic acids and derivatives, structure and physical properties of carboxylic acids and derivatives, reactivity of carboxylic acids and derivatives, nucleophilic acyl substitution, synthesis of acyl chlorides and anhydrides, reactions of carboxylic acid derivatives with organometallic reagents, complex metal hydride reduction.

Questions 1-6.

1. Assign each of the following into appropriate class of compounds.

2. Name the following compounds.

3. Arrange the following leaving groups in the order of increasing leaving ability.

   $-OCH_3$ $-Br$ $-N(CH_3)_2$ $-O^-$ $-OC(O)CH_3$

4. Arrange the following compounds in the order of increasing reactivities.

5. Why an acid cannot be converted into an ester in presence of a base catalyst?

241

6. Give examples for three different ways to prepare a tertiary alcohol (for example, triphenylmethanol shown here) from a Grignard reagent and a carbonyl compound (Hint: carbonate esters $(RO)_2C=O$ also undergo Grignard reaction). Indicate correct stochiometry for each reaction.

Answers to questions 1-6.

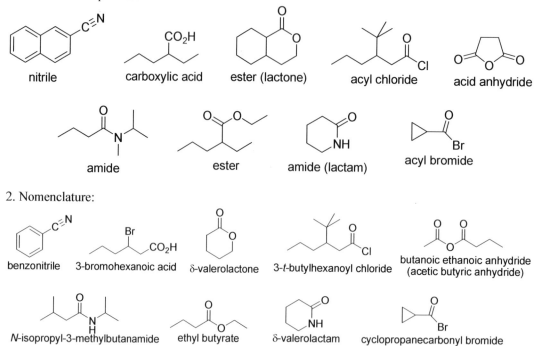

1. Classes of compounds: nitrile, carboxylic acid, ester (lactone), acyl chloride, acid anhydride, amide, ester, amide (lactam), acyl bromide

2. Nomenclature: benzonitrile, 3-bromohexanoic acid, δ-valerolactone, 3-t-butylhexanoyl chloride, butanoic ethanoic anhydride (acetic butyric anhydride), N-isopropyl-3-methylbutanamide, ethyl butyrate, δ-valerolactam, cyclopropanecarbonyl bromide

3. A good leaving group makes weak bond with carbon. Therefore, the weaker the bond to carbon the better the leaving group is. One way to predict leaving group ability is to consider their basicity. Weak bases are good leaving groups and strong bases are poor leaving groups. When comparing leaving groups, weaker bases are better leaving groups than stronger bases. Bromine makes weak bonds to carbon and is the best leaving group of those listed. Also, among the leaving groups listed, bromide ion is the weakest base. Nitrogen forms a strong bond with carbon. Once such bond is broken the resulting leaving group (amide anion) is a strong base. As a strong base it has a high tendency to re-form bond to the carbon. Therefore, amide is a poor leaving group. Remaining three leaving groups all contain oxygen. Of the three, the worst leaving group is the negatively charged oxygen. As a leaving group it would give an O^{2-} ion, which is an exceptionally strong base. Therefore, this is even worse leaving group than amide. Among the remaining two leaving groups, acetate ion is resonance stabilized and is a weaker base than methoxide ion. Therefore, it is a better leaving group, although not as good as bromide ion. Based on the above analysis the order of increasing leaving group ability is:

$$-O^- < -N(CH_3)_2 < -OCH_3 < -OC(O)CH_3 < -Br$$

An alternative approach is to consider conjugate acids of the leaving groups:

$HOCH_3$ HBr $HN(CH_3)_2$ HO^- $HOC(O)CH_3$

The order of increasing acid strength is:

$$HO^- < HN(CH_3)_2 < HOCH_3 < HOC(O)CH_3 < HBr$$

Conjugate base of the strongest acid is the weakest base and conjugate of the weakest acid the strongest base. Therefore, order of decreasing base strength and increasing leaving group ability is:

$$-O^- < -N(CH_3)_2 < -OCH_3 < -OC(O)CH_3 < -Br$$

4. Reactivities of carboxylic acids and their derivatives are related to the leaving group ability of the acyl substituent –Y:

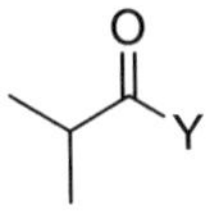

The better leaving group it is the more reactive the compound. Thus, one should compare leaving group abilities and arrange the compounds into appropriate order. Analysis of leaving group ability is given above (question 3). Therefore, the order of reactivities is:

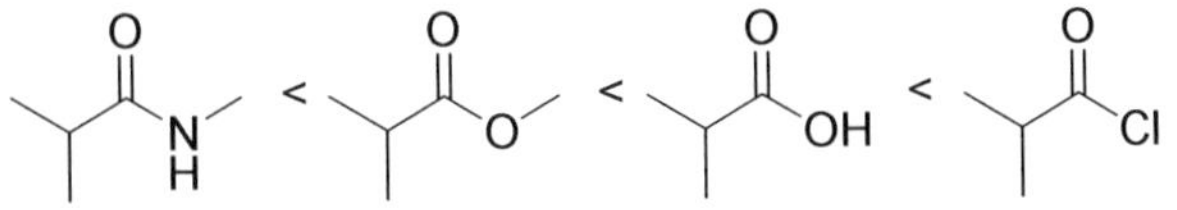

amide < anhydride < carboxylic acid < acyl chloride

5. Carboxylic acids and derivatives react by two general mechanisms:

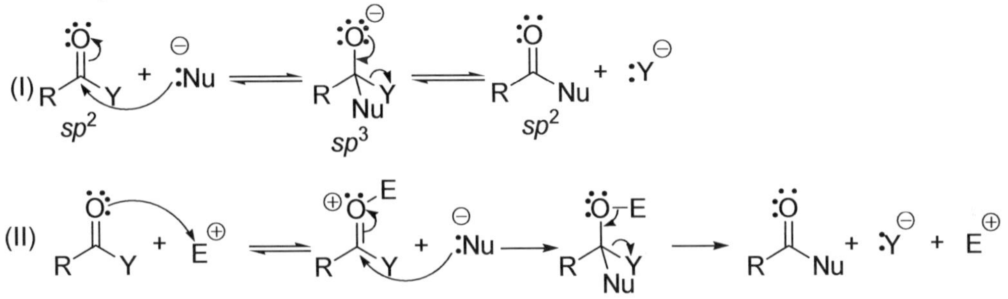

(I) Addition–elimination with nucleophile (sp² → sp³ → sp²)

(II) Acid-catalyzed addition–elimination

When applied to an esterification reaction the two mechanisms would be as follows:

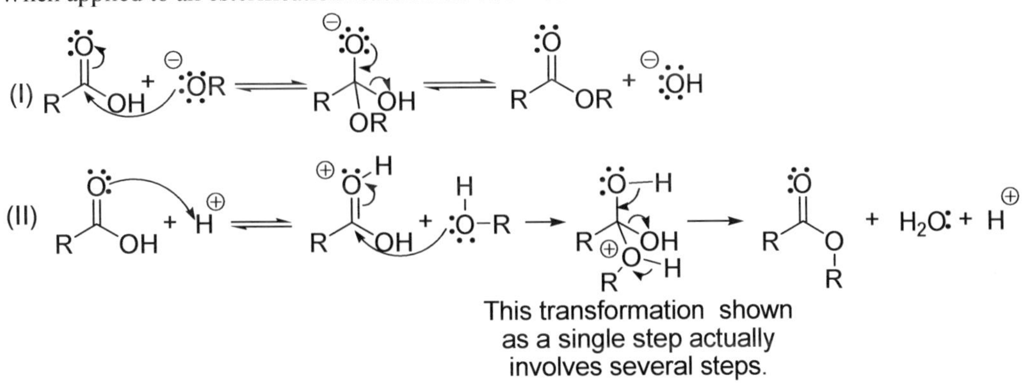

(I) Base-catalyzed pathway

(II) Acid-catalyzed pathway

This transformation shown as a single step actually involves several steps.

Examine the two mechanisms. There is a problem with the first mechanism. Can you spot it?

The mechanism involves a reaction between a carboxylic acid and a nucleophile, which is a base. Carboxylic acids are going to react with such species in an acid-base reaction and not in an acyl substitution reaction:

(I) R-C(=O)-O-H + :OR⁻ ⇌ R-C(=O)-O⁻ + ROH

The reaction product is a carboxylate ion, which is stable and does not undergo further reactions (negatively charged oxygen is a very poor leaving group – see question 3). Therefore, an ester cannot be prepared from a carboxylic acid under basic conditions. To prepare an ester from a carboxylic acid one is restricted to mechanism (II) and acidic reaction conditions. Esters can be prepared from bases (alkoxides) and other reactive carboxylic acid derivatives, such as chlorides and anhydrides, but not from carboxylic acids.

6. One way to prepare a tertiary alcohol is something you should already be familiar with – reaction of a ketone with a Grignard reagent:

benzophenone + 1) PhMgBr; 2) H₂O, H₃O⁺ → triphenylmethanol

Two equivalents of a Grignard reagent react with carboxylic acids derivatives that have a suitable leaving group (acyl, chlorides, acid anhydrides and esters) to give a tertiary alcohol. Therefore, there are three more methods to prepare triphenylmethanol:

Y-C(=O)-Y (Y = -Cl, -OR) + 1) 3 PhMgBr; 2) H₂O, H₃O⁺ → triphenylmethanol

Finally, carboxylic acid derivatives with two leaving groups such as phosgene, Cl-C(O)-Cl, as well as carbonates, RO-C(O)-OR, react with three equivalents of Grignard reagent to give tertiary alcohols. Therefore, reaction of three equivalents of phenylmagnesium bromide with either phosgene or a carbonate gives triphenylmethanol. Thus, there are two more methods to accomplish the desired transformation:

Ph-C(=O)-Y (Y = -Cl, -OR, -OC(O)R) + 1) 2 PhMgBr; 2) H₂O, H₃O⁺ → triphenylmethanol

Exercises in Organic Chemistry

Questions 7-9.

7. Provide formulas of the reaction products.

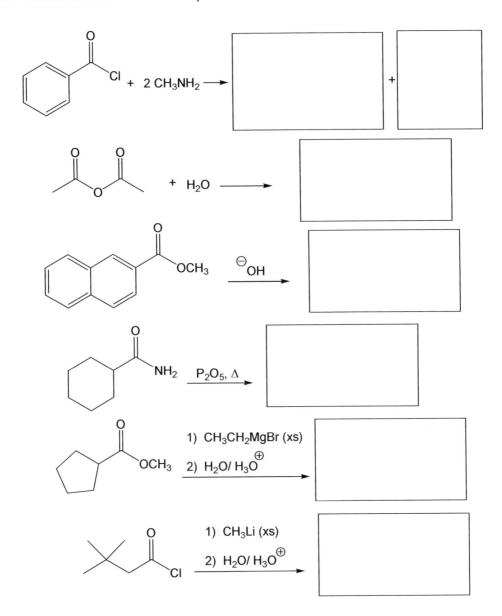

8. Provide formulas of the starting materials.

cyclopentane-COOH $\xrightarrow{SOCl_2}$ cyclopentane-C(O)Cl

? $\longrightarrow$ CH$_3$CH$_2$COO$^-$ K$^+$

? $\longrightarrow$ mixed anhydride of propanoic acid and benzoic acid

? $\xrightarrow{\text{1) LiAlH}_4\text{, THF} \quad \text{2) H}_2\text{O}}$ CH$_3$CH$_2$CH$_2$NHCH$_2$CH$_3$

9. Provide formulas of the reagents.

(CH$_3$)$_2$CHCOOH + [?] $\longrightarrow$ anhydride of isobutyric acid

CH$_3$COOH + CH$_3$NH$_2$ $\longrightarrow$ [?] $\longrightarrow$ [?] $\longrightarrow$ CH$_3$C(O)NHCH$_3$

Answers to questions 7-9.

7. Reaction products:

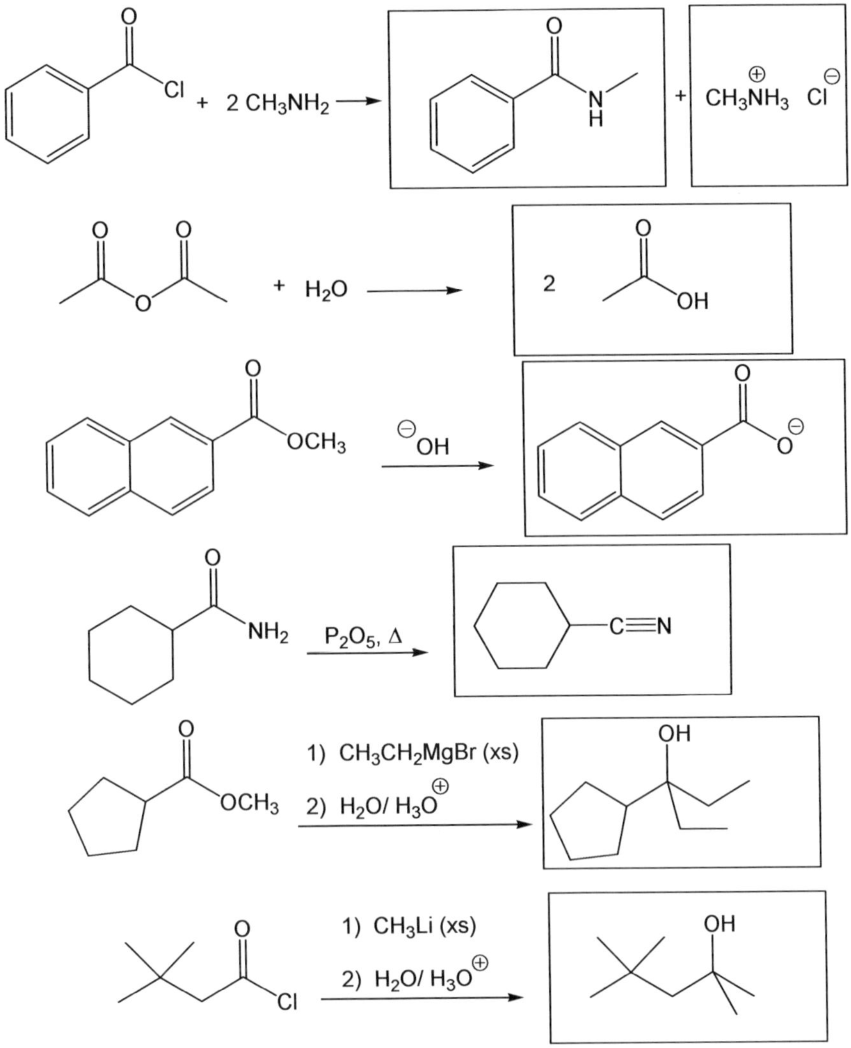

PhC(O)Cl + 2 CH₃NH₂ ⟶ PhC(O)NHCH₃ + CH₃NH₃⁺ Cl⁻

(CH₃CO)₂O + H₂O ⟶ 2 CH₃COOH

2-Naphthyl-C(O)OCH₃ + ⁻OH ⟶ 2-Naphthyl-C(O)O⁻

Cyclohexyl-C(O)NH₂ —P₂O₅, Δ→ Cyclohexyl-C≡N

Cyclopentyl-C(O)OCH₃ —1) CH₃CH₂MgBr (xs); 2) H₂O/H₃O⁺→ Cyclopentyl-C(OH)(CH₂CH₃)₂

(CH₃)₃CCH₂C(O)Cl —1) CH₃Li (xs); 2) H₂O/H₃O⁺→ (CH₃)₃CCH₂C(OH)(CH₃)₂

8. Starting materials:

(cyclopentanecarboxylic acid) + SOCl₂ → (cyclopentanecarbonyl chloride)

(benzoyl chloride) + (propanoate, O⁻ K⁺) → (mixed anhydride: propanoyl benzoyl anhydride)

(N-ethyl propanamide) or (N-ethyl acetamide) → 1) LiAlH₄, THF 2) H₂O → (N-ethyl propylamine)

9. Reagents:

(isobutyric acid) — P₂O₅, heat → (isobutyric anhydride)

(acetic acid) + CH₃NH₂ → [CH₃COO⁻ CH₃NH₃⁺] —heat→ (N-methylacetamide, CH₃CONHCH₃)

Additional Exercises: Questions 10-18

10. In each of the following compounds circle the leaving group.

$$H_3C-\overset{O}{\underset{}{C}}-OH \quad H_3C-\overset{O}{\underset{}{C}}-Cl \quad H_3C-\overset{O}{\underset{}{C}}-NH_2 \quad H_3C-\overset{O}{\underset{}{C}}-O-\overset{O}{\underset{}{C}}-CH_3 \quad H_3C-\overset{O}{\underset{}{C}}-OCH_3$$

11. Provide formulas of the reagents.

12. Complete the following table giving the structure or name for each molecule listed.

Name	Structural formula
	propanoic anhydride structure (CH₃CH₂C(O)OC(O)CH₂CH₃)
ethanoic methanoic anhydride	
	CH₃CH₂C(O)N(CH₃)CH₂CH₃ (N-ethyl-N-methylpropanamide)
2-chloropropanenitrile	
	2-pyrrolidinone (γ-butyrolactam)
3,3-dimethylhexanamide	
	ethyl benzoate
propyl 2-oxobutanoate	
	phenyl acetate
cis-3-propylcyclopentanecarboxylic acid	

Exercises in Organic Chemistry

13. Provide products of the following reactions.

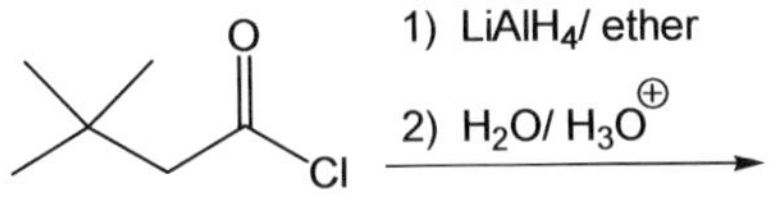

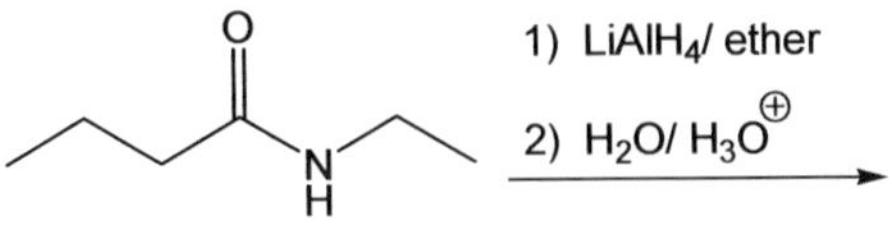

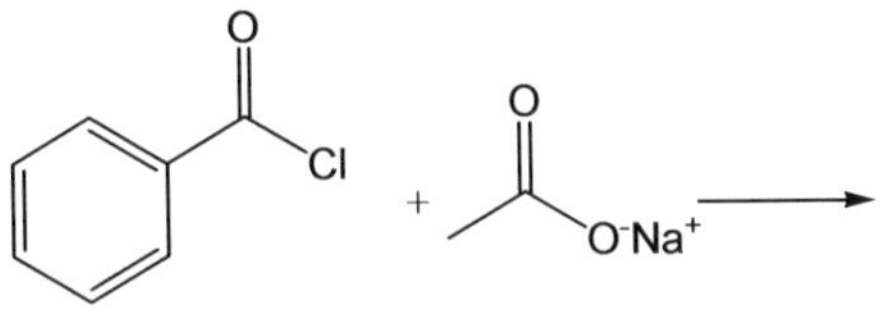

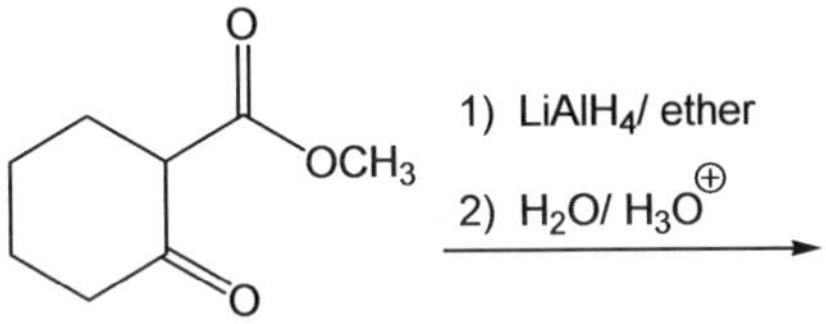

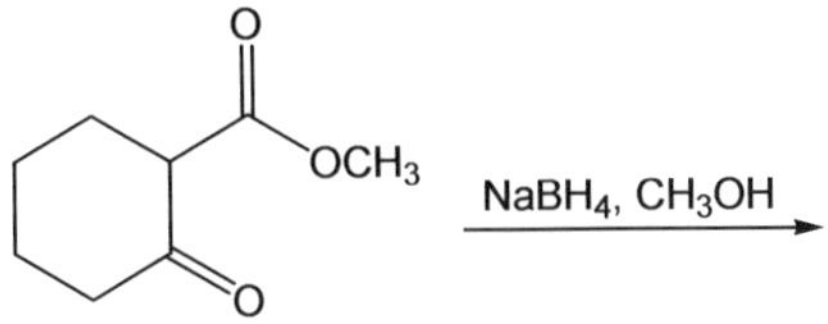

(CH₃)₃C-CH₂-C(=O)-Cl
1) LiAlH₄/ ether
2) H₂O/ H₃O⁺

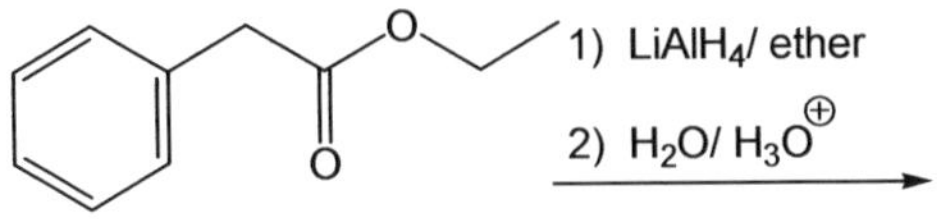

Ph-CH₂-C(=O)-O-CH₂-CH=CH-... (phenylacetic acid allyl-type ester)
1) LiAlH₄/ ether
2) H₂O/ H₃O⁺

CH₃CH₂CH₂-C(=O)-NH-CH₂CH₃
1) LiAlH₄/ ether
2) H₂O/ H₃O⁺

Ph-C(=O)-Cl + CH₃-C(=O)-O⁻Na⁺ ⟶

(2-oxocyclohexyl)-C(=O)-OCH₃
1) LiAlH₄/ ether
2) H₂O/ H₃O⁺

(2-oxocyclohexyl)-C(=O)-OCH₃
NaBH₄, CH₃OH ⟶

14. A product of the following reaction is:

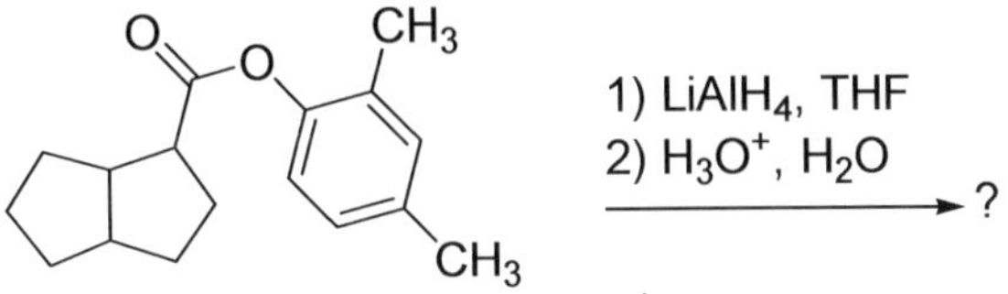

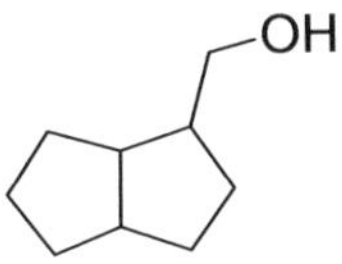

bicyclic-C(=O)-O-(2,4-dimethylphenyl)
1) LiAlH₄, THF
2) H₃O⁺, H₂O
⟶ ?

a) bicyclic-CH₂-OH

b) H₃C—(phenyl)—CH₃

c) bicyclic-CH₂-OH

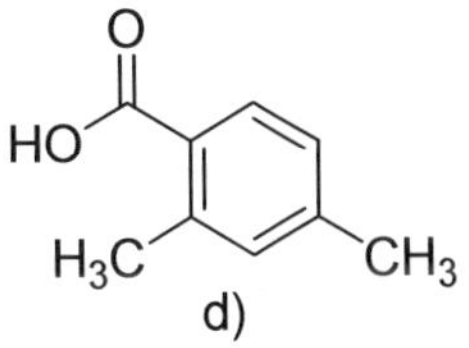

d) HO-C(=O)—(phenyl with H₃C and CH₃)

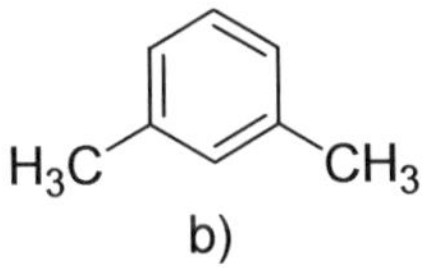

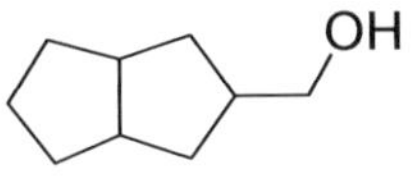

251

15. Which class of compounds is the least reactive?

    a) Carboxylic acids
    b) Amides
    c) Acyl chlorides
    d) Esters

16. Which of the following compounds is the most stable?

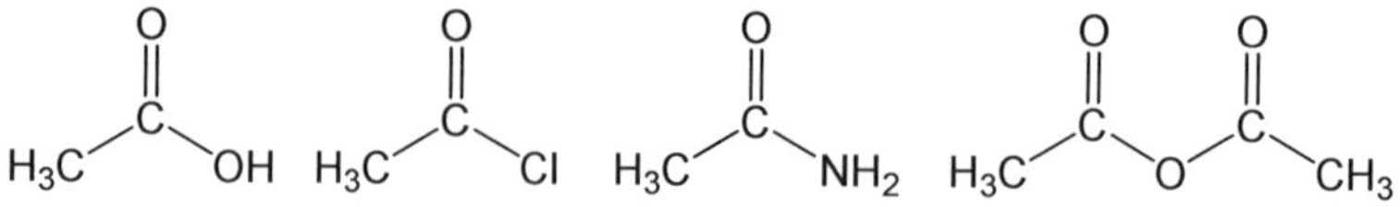

    a) $H_3C-C(=O)-OH$
    b) $H_3C-C(=O)-Cl$
    c) $H_3C-C(=O)-NH_2$
    d) $H_3C-C(=O)-O-C(=O)-CH_3$

17. In the following reaction pyridine is a:

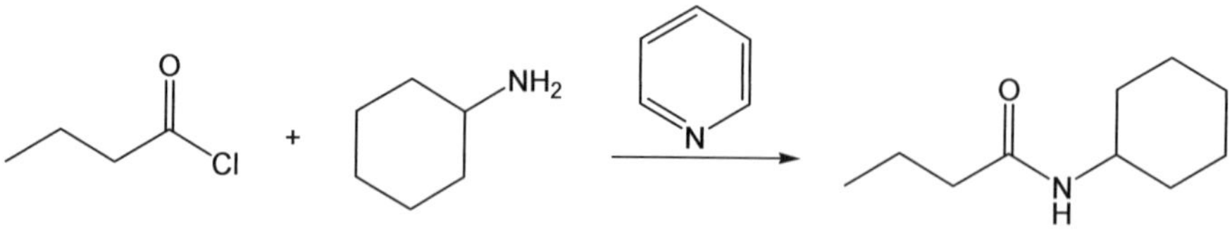

    a) nucleophile
    b) base
    c) catalyst
    d) leaving group

18. The fact that in the following hydrolysis of an ester, the reaction proceeds without a rearrangement (the alkyl group of the ester gives unrearranged alcohol) is an indication that the reaction mechanism is not:

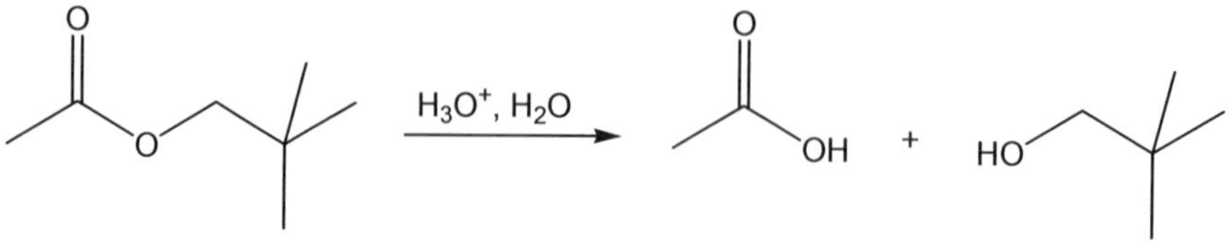

    a) S_N1
    b) S_N2
    c) Addition/Elimination
    d) Elimination/Addition

Answers to Additional Exercises: Questions 10-18

10. Leaving groups:

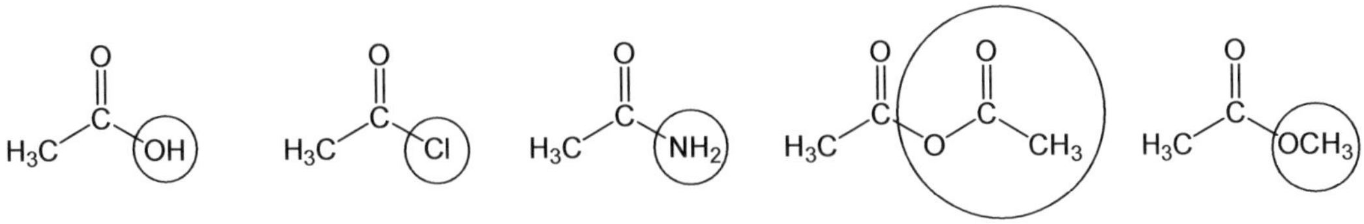

The leaving groups circled are: OH, Cl, NH₂, the anhydride O-C(O)CH₃ group, and OCH₃ on the respective acetyl derivatives (H₃C-CO-X).

11. Reagents:

Reaction 1: ester (3,3-dimethylbutanoic acid ethyl ester) + (iBu)₂AlH, THF, -78°C → aldehyde (3,3-dimethylbutanal)

Reaction 2: isobutyl cyanide (R-C≡N) + H₃O⁺, H₂O, heat or ⁻OH, H₂O, heat → carboxylic acid (3-methylbutanoic acid)

Reaction 3: cyclobutanecarboxylic acid + P₂O₅, heat → cyclobutanecarboxylic anhydride

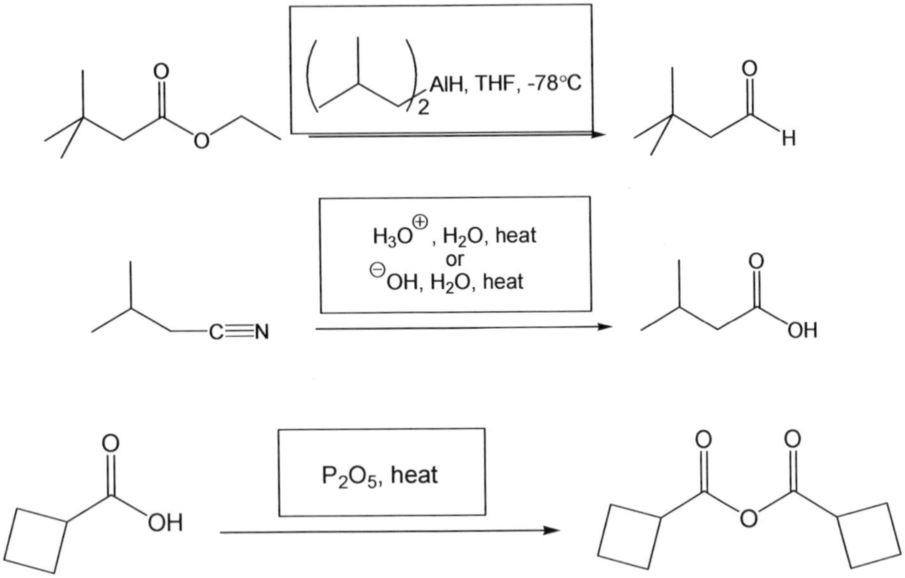

12. Table:

Name	Structural formula
propanoic anhydride or propionic anhydride	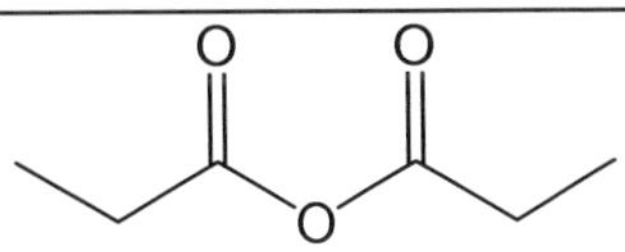 CH₃CH₂C(O)-O-C(O)CH₂CH₃
ethanoic methanoic anhydride	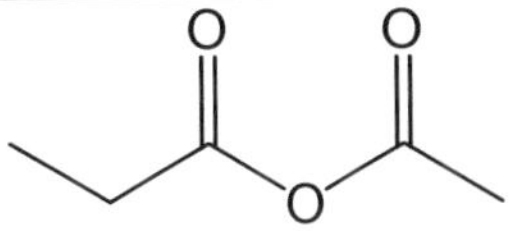 CH₃CH₂C(O)-O-C(O)CH₃
N-ethyl-N-methylpropanamide	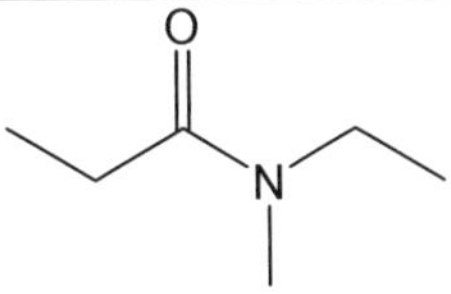 CH₃CH₂C(O)-N(CH₃)(CH₂CH₃)

2-chloropropanenitrile	
γ-butyrolactam	
3,3-dimethylhexanamide	
ethyl benzoate	
propyl 2-oxobutanoate	
phenyl ethanoate or phenyl acetate	
cis-3-propylcyclopentanecarboxylic acid	

13. Reaction products:

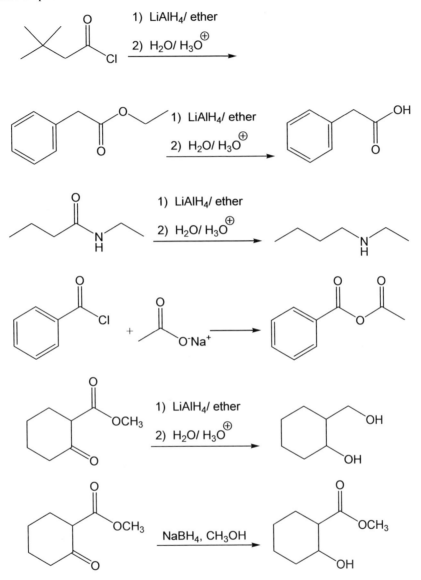

14. a); 15. b); 16. c); 17. b); 18. a).

22. Carbonyl Compounds III: Substitution α to Carbonyl Group

Concepts to review: acidity of α-hydrogens, keto-enol tautomerism, halogenation α to carbonyl groups, kinetic vs. thermodynamic deprotonation, Michael addition, aldol reaction and condensation, Claisen condensation, intramolecular reactions, acetoacetic and malonic ester syntheses.

Questions 1-7.

1. Which one is the strongest acid among the following molecules:

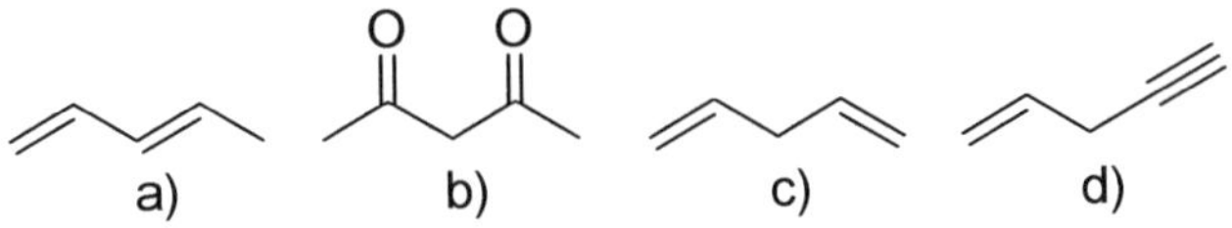

O
O
a)
b)
c)
d)

2. Which of the following compounds is the strongest acid?

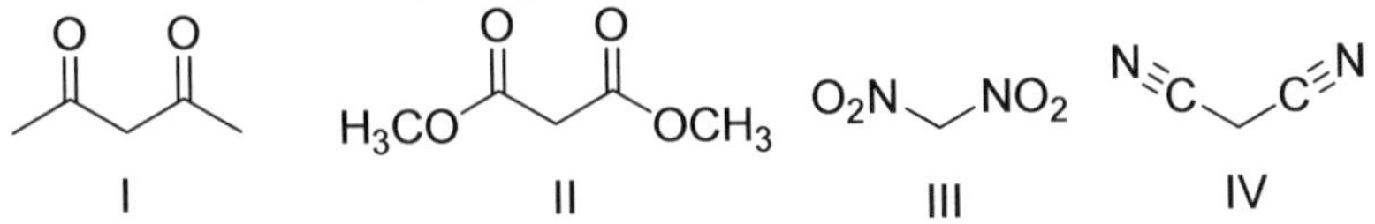

O
O
I
O
O
H3CO
OCH3
II
O2N
NO2
III
N≡C
C≡N
IV

3. Which of the following protons is the most acidic.

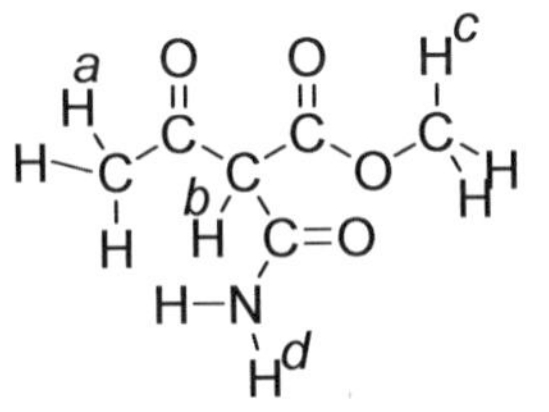

a
b
c
d
H
C=O
H–N

4. Provide the formulas of the products.

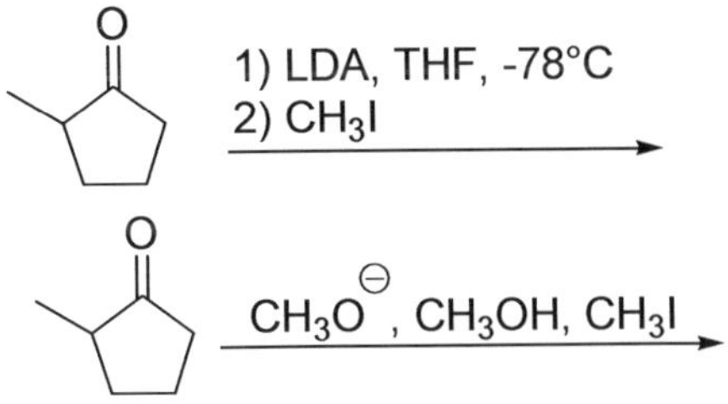

1) LDA, THF, -78°C
2) CH3I
CH3O, CH3OH, CH3I

5. Provide formula of the reaction product.

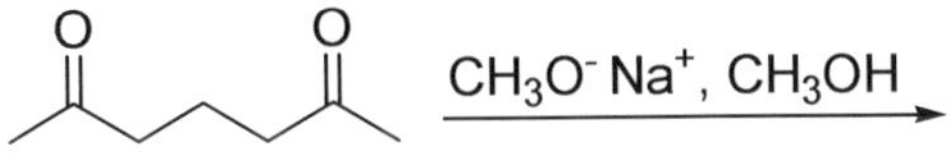

CH3O- Na+, CH3OH

6. How would you prepare the following compound from organic compounds that contain three carbon atoms or less and inorganic reagents?

$$\text{CH}_3\text{CH(CHO)CH(OH)CH}_3$$

7. How would you prepare the following compound starting with organic compounds that contain four carbon atoms or less and inorganic reagents?

(compound: 2-methyl-2-hexenal, CHO on carbon bearing methyl, C=C, then CH₂CH₃)

Answers to questions 1-7.

1. Acidity of a hydrogen bonded to carbon is very low. Hydrogens in ordinary alkanes have pK_a values of ~50. However, presence of a neighboring electronegative substituent considerably increases acidity of hydrogens through the substituents negative inductive effect and possible stabilization of the resulting conjugate base by resonance. Therefore, substituent with the strongest -I effect is going to produce the strongest acid. Among the compounds listed above, oxygen atom and, hence, the carbonyl group, have the strongest negative inductive effect. Inductive effects of an sp and sp^2 hybridized carbon atoms are considerably smaller. Thus, the strongest acid is the compound b).

2. This question is similar to the previous one. Now you have to compare inductive effects of a number of oxygen- and nitrogen-containing substituents. Order of their inductive effects is: $-CN < -CO_2CH_3 < -COCH_3 < -NO_2$. Therefore, compound III is the strongest acid as it has two of the substituents with the strongest electron withdrawing effect.

3. Substituent effects are additive. Proton a) is activated by only one carbonyl substituent. Although it is acidic, it is not as acidic as a proton that is on the carbon next to more than one electron-withdrawing substituent. Proton b) is on the carbon with three electron-withdrawing substituents and it is the most acidic of the four. Protons c) and d) are not considered to be acidic.

4. Both of the reactions involve alkylation of a carbonyl compound. The first one is carried out with LDA (lithium diisopropylamide) as a base. That reagent generates a kinetic (less substituted) enolate anion and the alkylaiton occurs on the less substituted α carbon. In the second reaction, the base is methoxide ion, which generates thermodynamic (more substituted) enolate anion. Therefore, the substitution occurs at the more substituted α carbon. In the latter case a considerable amount of polyalkylation products is also obtained.

2-methylcyclopentanone $\xrightarrow{\text{1) LDA, THF, -78°C} \\ \text{2) CH}_3\text{I}}$ 2,5-dimethylcyclopentanone

2-methylcyclopentanone $\xrightarrow{CH_3O^{\ominus}, CH_3OH, CH_3I}$ 2,2-dimethylcyclopentanone

5. The reaction in an intramolecular aldol reaction and the product is a cyclic compound. An intramolecular aldol reaction can give two possible cyclic products depending on the number of atoms between the two carbonyl groups. Five and six member rings are produced preferentially to 3, 4, 7, 8 - member rings because of the ring strain. In this case there are two possible products – one with a five-member ring and the other one with a three-member ring. The only reaction product is going to be the one with a five-member ring.

6. A question like this may be challenging to some students. First consider the starting material – it has six carbon atoms and two functional groups. You have to start with carbon compounds that have no more than three carbon atoms. Also, consider the actual functional groups – the molecule is a β-hydroxyaldehyde. A method for preparation of β-hydroxyaldehyde is an aldol reaction. In aldol reaction two molecules of an aldehyde react to give the product that is a dimer – has twice the number of atoms of the starting material. Therefore, aldol reaction of an aldehyde with three carbon atoms should give the product with six carbon atoms which is also a β-hydroxyaldehyde:
Thus, the correct starting material is propanal and the only inorganic reagent that is needed is sodium hydroxide (or another suitable base).

7. This question is similar to the previous one. The only difference is that now you have to recognize that the reaction is an aldol condensation, which gives an α,β-enone. Therefore, the starting material is butanal, inorganic reagent is again sodium hydroxide and the reaction has to be done with heating to effect elimination of water from the intermediate aldol reaction product.

Questions 8-11.

8. Provide formulas of the products of the following reactions.

2-acetylnaphthalene + 3Cl$_2$ $\xrightarrow{OH^-}$

cyclopentanone + Br$_2$ $\xrightarrow{H_3O^+}$

2 CH$_3$CHO $\xrightarrow{OH^-, \Delta}$

pentane-2,4-dione + but-2-enal $\xrightarrow{OH^-}$

ethyl 3-oxobutanoate (ethyl acetoacetate) $\xrightarrow{\text{1) CH}_3\text{CH}_2\text{O}^-, \text{CH}_3\text{CH}_2\text{OH} \quad \text{2) CH}_3\text{I} \quad \text{3) H}_3\text{O}^+, \text{H}_2\text{O}, \Delta}$

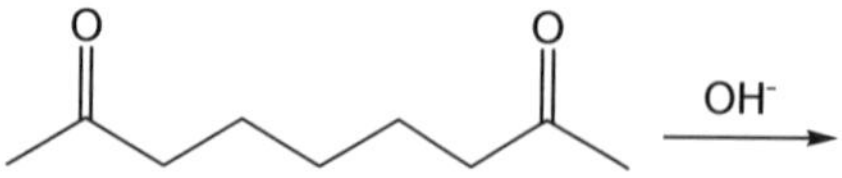

heptane-2,6-dione $\xrightarrow{OH^-}$

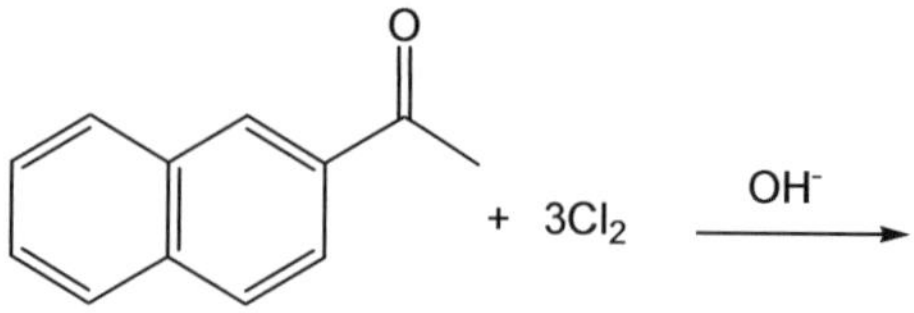

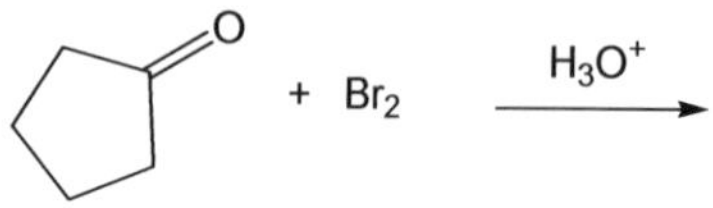

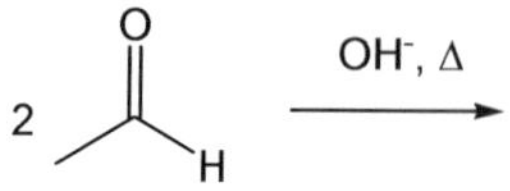

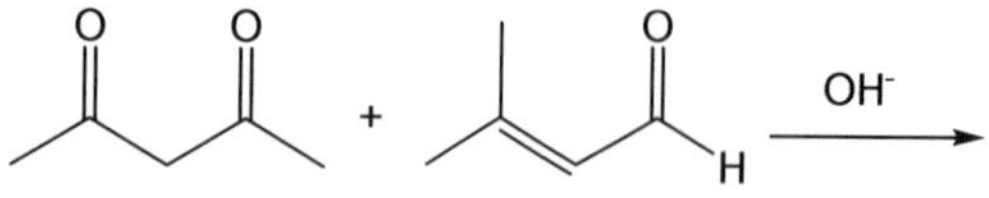

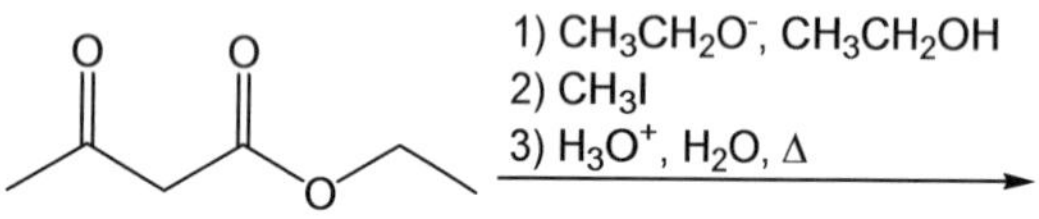

9. Provide the final product of the following reaction sequence:

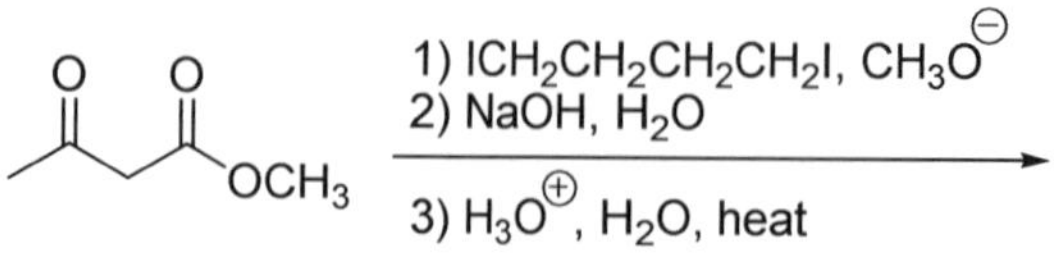

methyl 3-oxobutanoate (OCH$_3$ ester) $\xrightarrow{\text{1) ICH}_2\text{CH}_2\text{CH}_2\text{CH}_2\text{I, CH}_3\text{O}^{\ominus} \quad \text{2) NaOH, H}_2\text{O} \quad \text{3) H}_3\text{O}^{\oplus}, \text{H}_2\text{O, heat}}$

10. Provide reagents for the following reactions:

pivaldehyde $\xrightarrow{\boxed{}}$ 2,2-dimethyl-3-hydroxy... OH + CHCl$_3$

pentane-2,4-dione + ethyl acrylate $\xrightarrow{\boxed{}}$ cyclic product

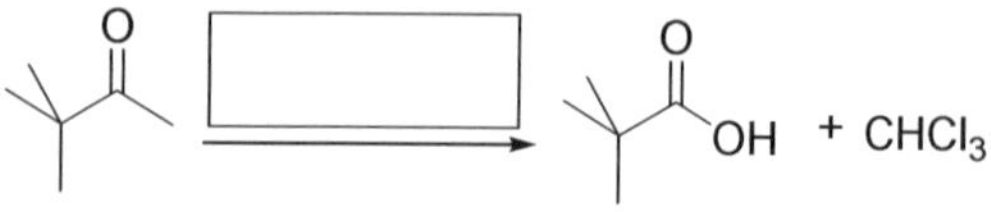

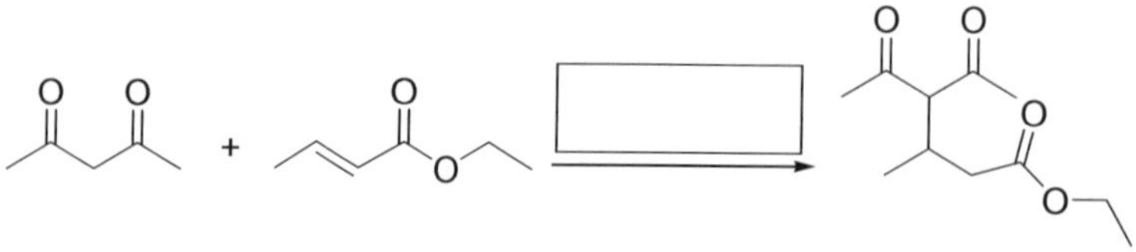

Exercises in Organic Chemistry

11. Provide starting materials for the following reactions:

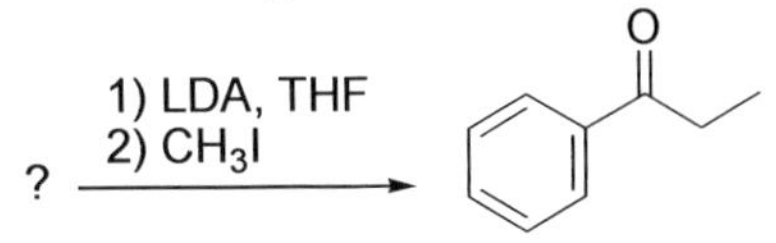

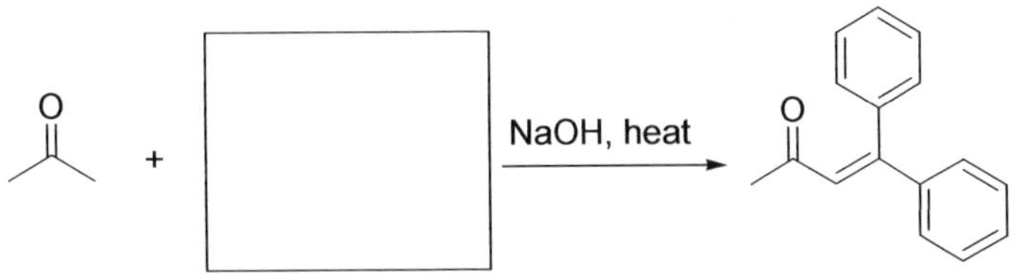

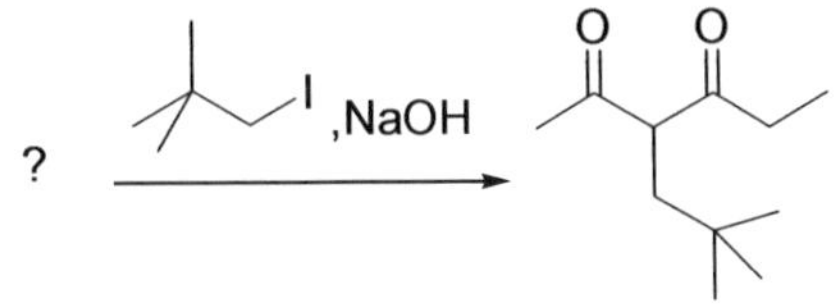

? $\xrightarrow{\text{1) LDA, THF} \\ \text{2) CH}_3\text{I}}$ propiophenone

acetone + [ ] $\xrightarrow{\text{NaOH, heat}}$ PhC=CH–C(O)–CH₃ (1,1-diphenyl enone)

? $\xrightarrow{\text{neopentyl-I, NaOH}}$ alkylated β-diketone

Answers to questions 8-11.

8. Reaction products:

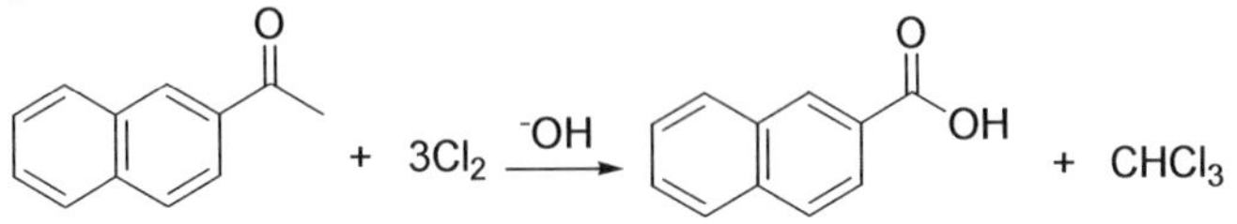

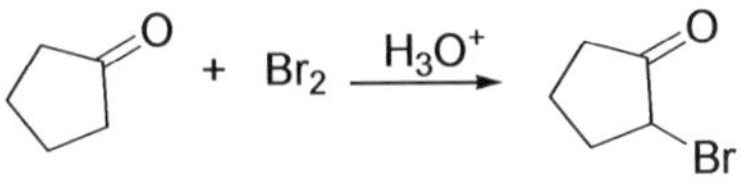

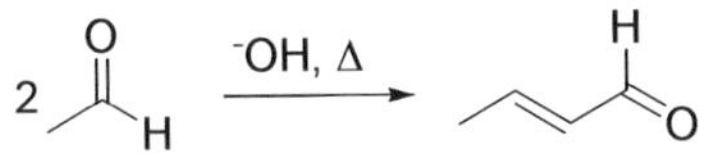

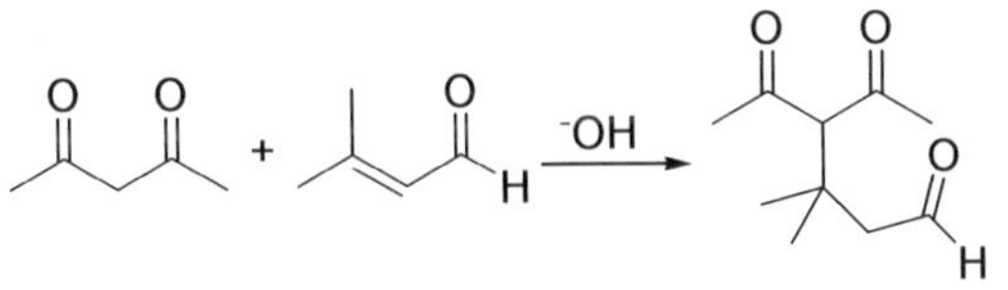

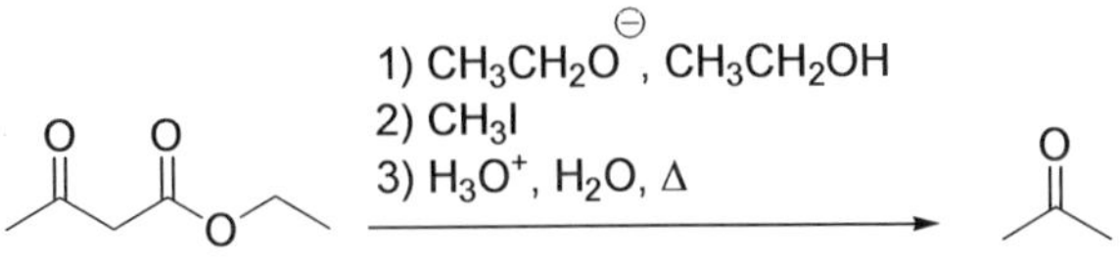

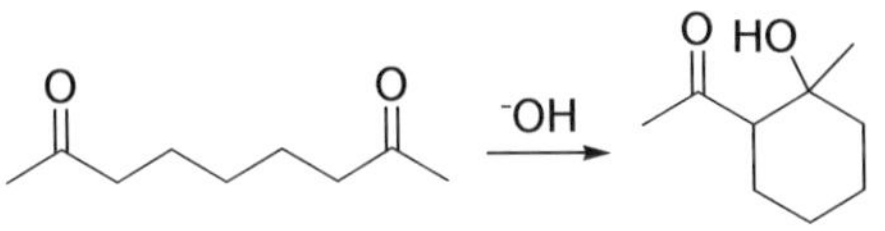

2-acetylnaphthalene + 3Cl₂ $\xrightarrow{\text{¯OH}}$ 2-naphthoic acid + CHCl₃

cyclopentanone + Br₂ $\xrightarrow{\text{H}_3\text{O}^+}$ 2-bromocyclopentanone

2 CH₃CHO $\xrightarrow{\text{¯OH, }\Delta}$ 2-butenal (crotonaldehyde)

pentane-2,4-dione + CH₃–CH=CH–CHO $\xrightarrow{\text{¯OH}}$ Michael adduct

ethyl acetoacetate $\xrightarrow{\text{1) CH}_3\text{CH}_2\text{O}^\ominus, \text{CH}_3\text{CH}_2\text{OH} \\ \text{2) CH}_3\text{I} \\ \text{3) H}_3\text{O}^+, \text{H}_2\text{O, }\Delta}$ 2-butanone

heptane-2,6-dione $\xrightarrow{\text{¯OH}}$ 1-acetyl-2-hydroxy-2-methylcyclohexane (intramolecular aldol)

261

9. The reaction sequence is called acetoacetic ester synthesis. It involves an alkylation of an acetoacetic ester followed by decarboxylation. Final reaction product is an alkylated ketone. In this particular reaction, the reagent is actually an alkyl dihalide and it can react twice in an alkylation reaction to give a ring compound. Thus, result of the alkylation reaction is formation of a five-member ring. In the final step, the ester group is hydrolyzed and decarboxylated to give the final product cyclopentyl methyl ketone.

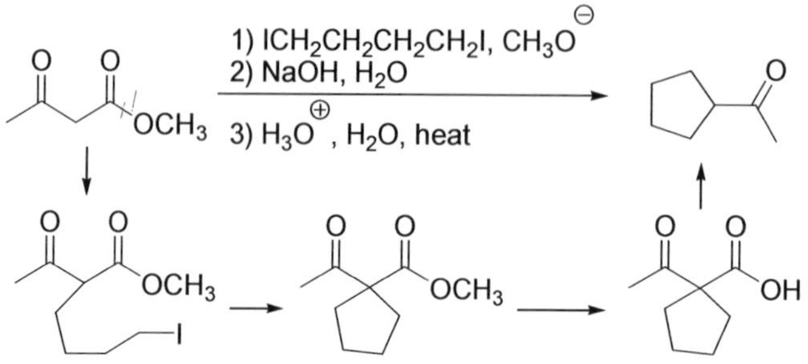

10. Reagents: Cl$_2$, NaOH

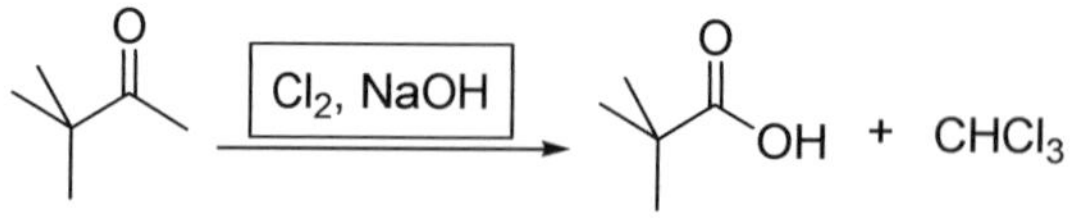

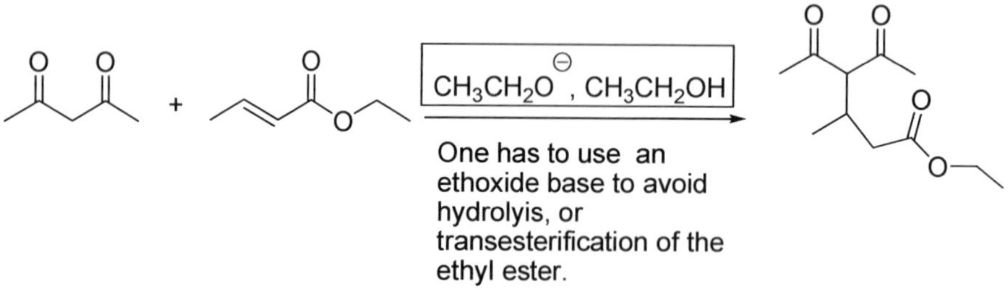

11. Starting materials:

1) LDA, THF
2) CH$_3$I

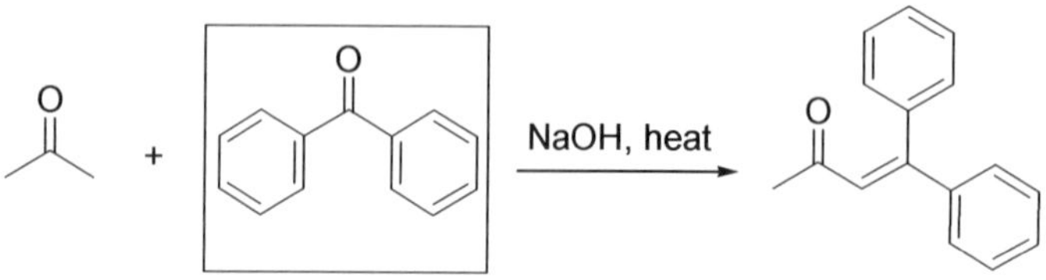

NaOH, heat

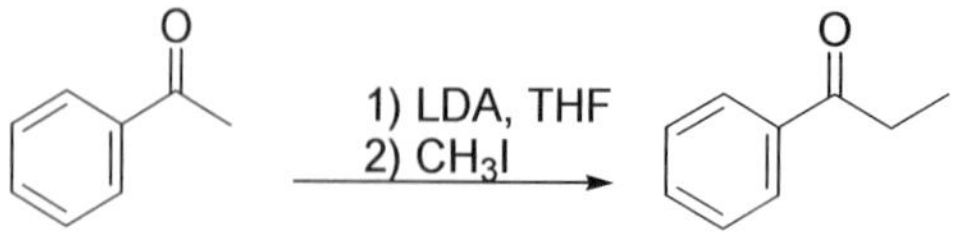

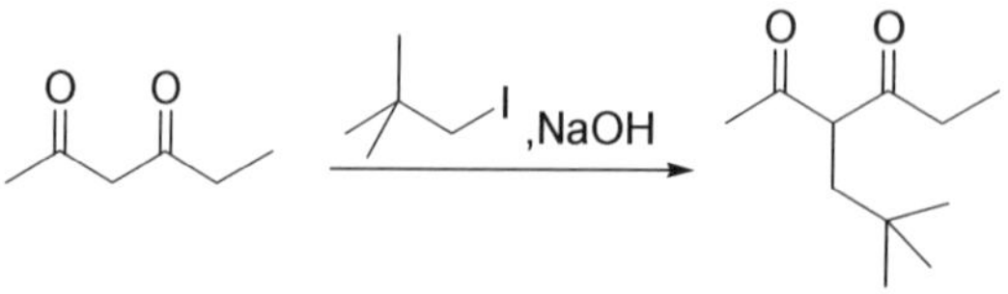

Additional Exercises: Questions 12-16

12. Provide formulas of the products of the following reactions.

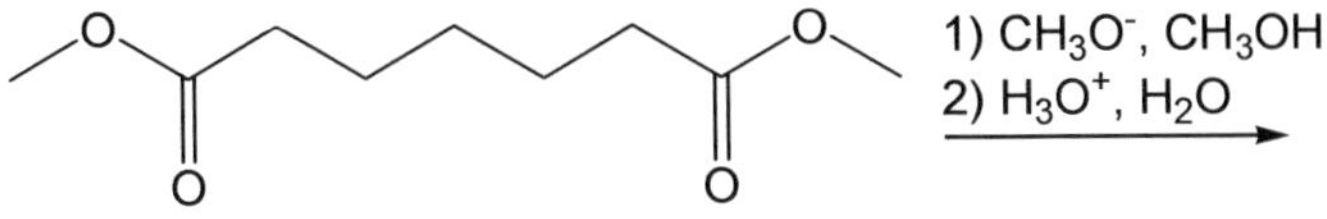
1) CH3O-, CH3OH
2) H3O+, H2O

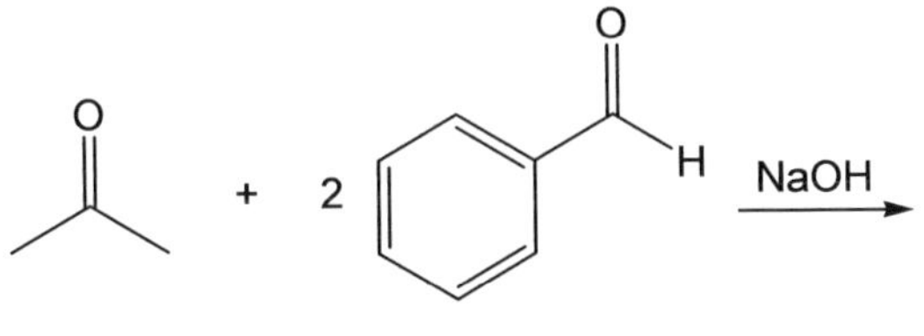
+ 2
H
NaOH

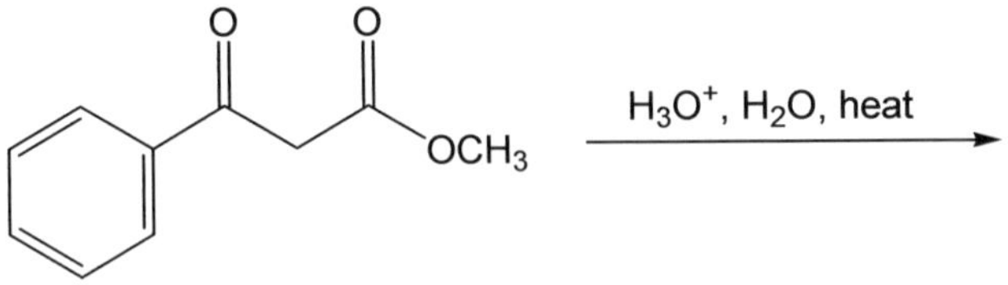
OCH3
H3O+, H2O, heat

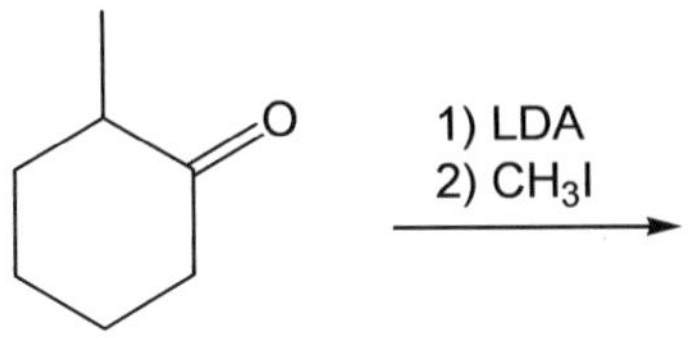
1) LDA
2) CH3I

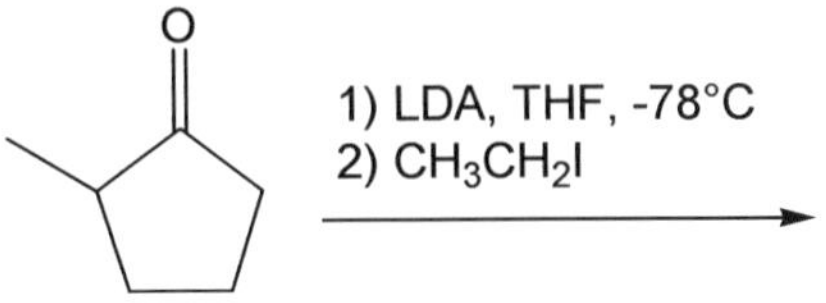
1) LDA, THF, -78°C
2) CH3CH2I

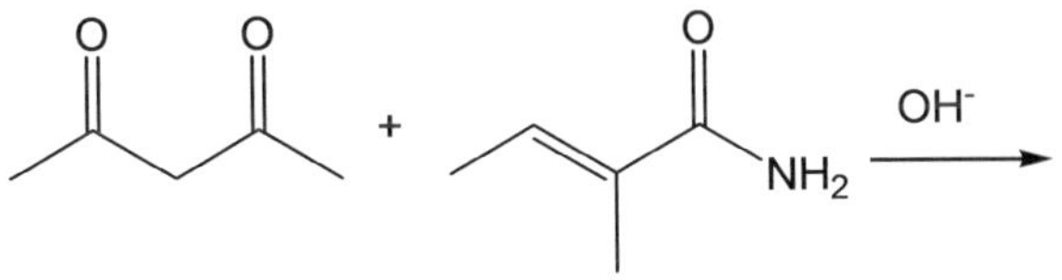
+
NH2
OH-

13. Provide the structures of missing compounds:

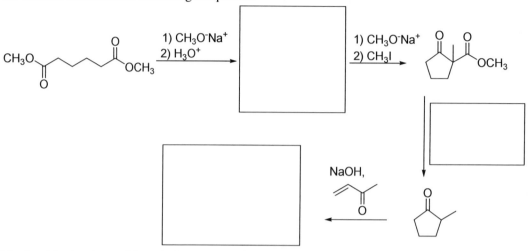

14. Provide structures of the missing compounds:

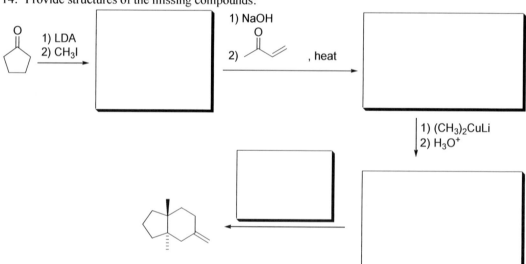

15. Robinson annulation is:

   a) a method for formation of six-member rings.
   b) a method for formation of four-member rings.
   c) an intramolecular Claisen condensation.
   d) an intramolecular crossed Claisen condensation.

16. What is the product of the following reaction?

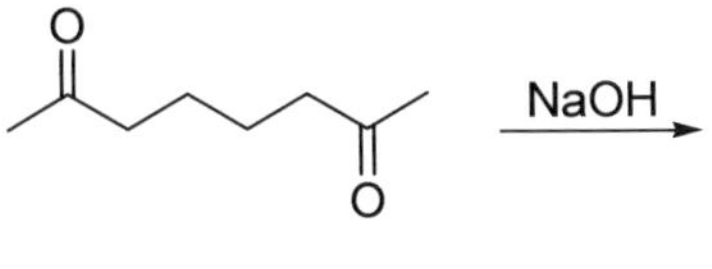

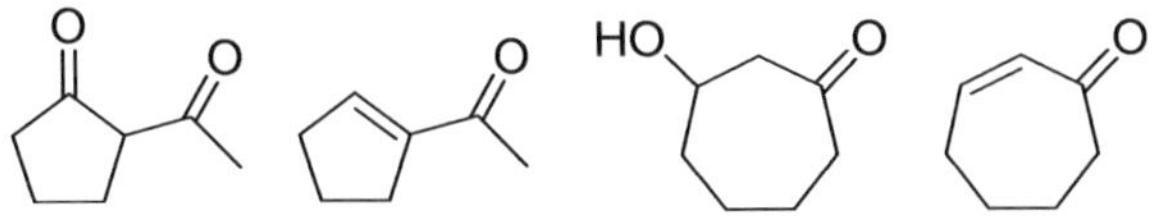

a) b) c) d)

Answers to Additional Exercises: Questions 12-16

12. Reaction products:

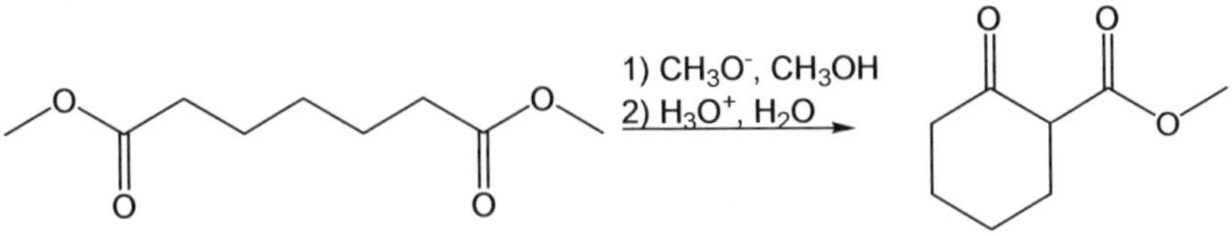

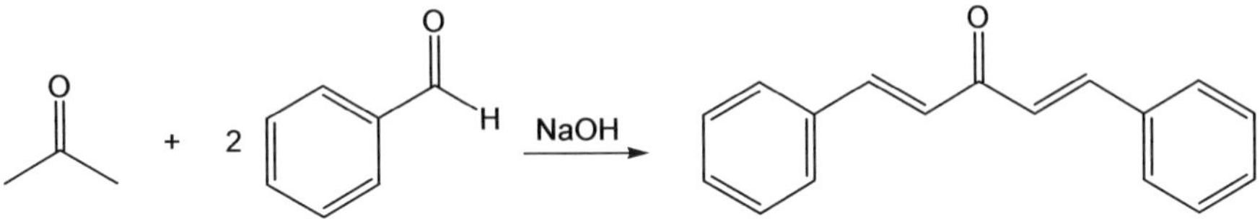

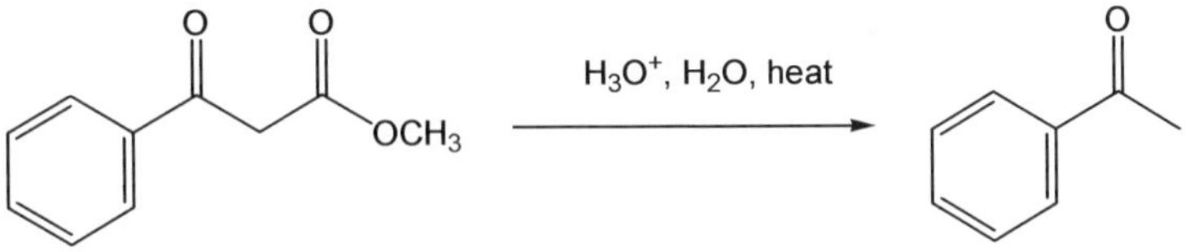

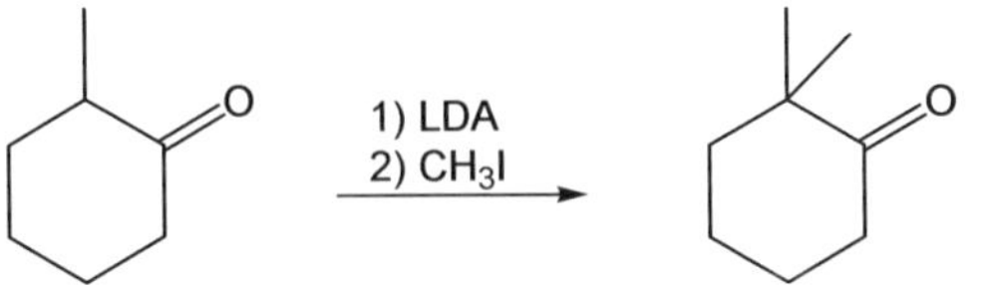

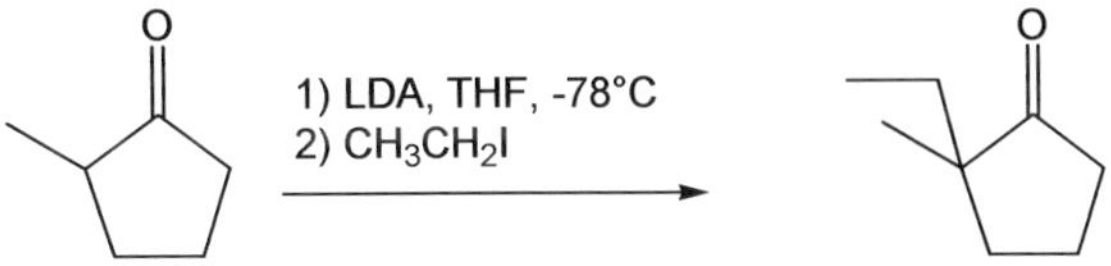

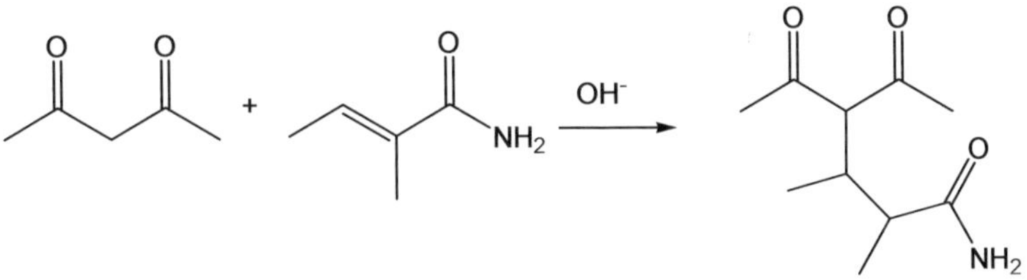

13. Formulas of the missing compounds:

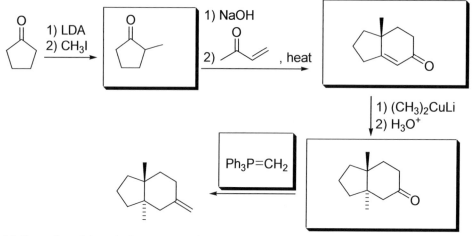

14. Formulas of the missing compounds:

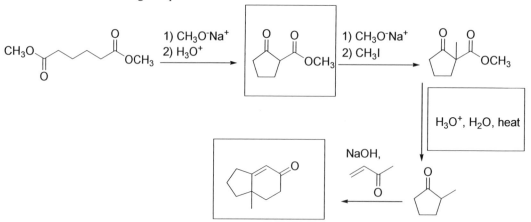

15. a); 16. b).

Exercises in Organic Chemistry

23. Polymers

Concepts to review: monomers, mechanisms of polymerization, addition and condensation polymers, common starting materials for preparation of polymers.

1. Identify the smallest repeating unit in each of the following polymers?

   a) [polymer chain with NO₂ groups on alternating carbons]

   b) [polyether chain with -CH₂CH₂-O- repeat units]

   c) [polyester chain with ethylene glycol and malonate units]

2. What monomers (starting materials) would you use to prepare each of the following polymers?

   a) [polymer with alternating double bonds]

   b) [polymer chain with OAc groups on alternating carbons]

   c) [polyester with -(CH₂)₃-C(O)-O-(CH₂)₃- repeat]

3. Draw structure of a polymer that is produced by polymerization of each of the following monomers.

   a) HO–C₆H₄–C(O)OH   b) CH₂=CH–C≡N   c) β-propiolactone

4. What is a suitable starting material for preparation of a polyurethane?

   a) an isocyanate.
   b) a carbamate.
   c) an amide.
   d) a carbamic acid.

5. Which of the following compounds can be used as a cross-linking agent in preparation of a polyester?

a) HO–C(=O)–CH₂–C(=O)–OH   b) β-propiolactone   c) HO–CH₂CH₂CH₂–OH   d) HOCH₂–C(CH₂OH)₂–CH₂OH (structure showing three –OH groups)

Answers.

1. It is relatively easy to identify self-repeating units in the first two polymers. They are made from a single monomer. The third one is a condensation polymer made from two monomers and it is somewhat more difficult to identify the repeating unit. The main "trick" here is to identify really the smallest repeating unit (not a dimer) and to include all the atoms in it (in the example c) students frequently either do not include the end ester oxygen, or include both ester oxygens).

a) polymer with repeating –CH(NO₂)–CH₂– units

b) polymer with repeating –O–CH₂–CH₂– (polyether) units

c) polyester repeating unit –O–CH₂–CH₂–O–C(=O)–CH₂–C(=O)– type structure

2. Do not confuse this problem with the previous one. To answer this question you should 1) identify the smallest repeating unit, 2) figure out type of polymerization (addition or condensation) and 3) propose structure of the monomer(s). Thus, solutions are:
a) The smallest repeating unit is –CH=CH–. The polymer is an addition polymer, which means that, in the course of polymerization a π bond was converted into a σ bond. Therefore, we should remove the two "dangling" σ bonds and replace them by a π bond between the two carbon atoms, which gives us ethyne (acetylene), HC≡CH, as the correct answer.
b) The smallest repeating unit is –CH–C(OAc)–. Again the polymer is an addition polymer and by application of the previous procedure we get the answer: CH=C(OAc) or vinyl acetate.
c) In this case the smallest repeating unit is –O–CH₂–CH₂–CH₂–CH₂–C(O)–. One can alternatively identify the smallest repeating unit as: –CH₂–CH₂–CH₂–CH₂–CO₂–. Either way the starting polymer is a polyester and the mechanism of polymerization is a condensation reaction. A polyester can be made in a reaction between a diacid and a diol, or by polymerization of bifunctional molecule that contains both a hydroxyl and carboxyl groups. From the structure of the smallest repeating unit, it is clear that a bifunctional compound was the starting material. To identify the starting material, we have to replace the "dangling bonds" with H and OH groups to get carboxyl and hydroxyl groups. Thus, the staring material is HO–CH₂–CH₂–CH₂–CH₂–CO₂–H. Alternatively, another correct answer is that the staring material is the

following lactone, which, in a base catalyzed polymerization, gives the same polyester:

3. Now the process is reversed. We already have the starting monomer and now we should identify the type of polymerization, which in turn should allow us to predict the structure of the polymer.
a) This is a bifunctional compound: a hydroxyester. It undergoes condensation polymerization. Therefore, one should convert the end hydroxyl and carboxyl groups into ester groups:

b) The compound is an alkene and undergoes an addition polymerization (don't be confused by -C≡N, it is only a substituent). To predict structure of the polymer, one should convert the *p* into *s* bond and generate polymer chain:

c) This monomer is a lactone and the "trick" here is to recognize that, in a base catalyzed polymerization, lactone ring opens to give a hydroxyacid – in this case 3-hydroxypropanoic acid. Next, we should generate a polymer chain by converting hydroxyl and carboxyl groups into ester groups:

4. a); 5. d) – it is the only starting material that would not produce a liner polymer.

Exercises in Organic Chemistry

Test 6: Chapters 20-23

1. Provide a mechanism for NaBH₄ reduction of an aldehyde.

2. How would you prepare the following compound?

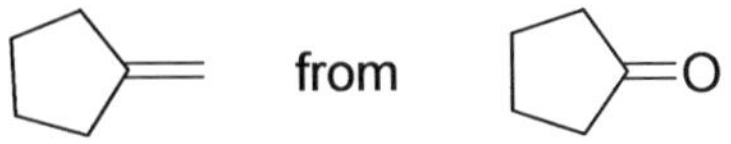

   cyclopentylidene (=CH₂) from cyclopentanone (=O)

3. Provide products of the following reactions.

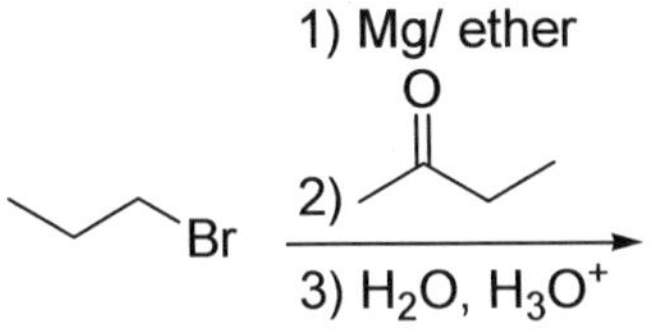

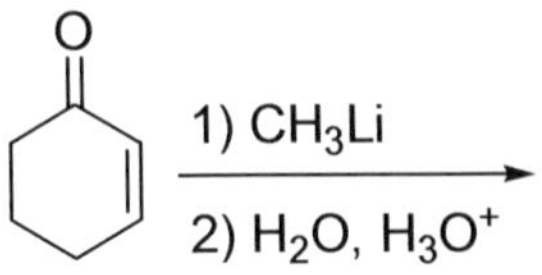

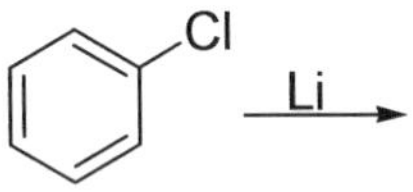

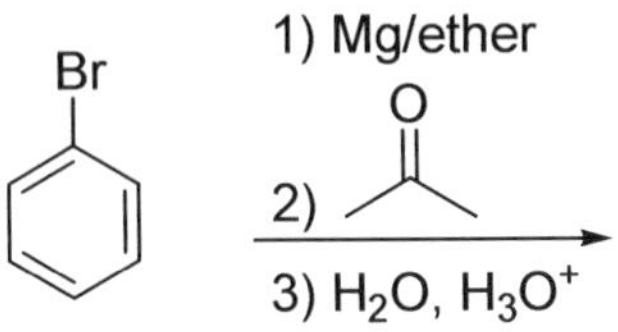

   CH₃CH₂CH₂Br → 1) Mg/ ether 2) CH₃CH₂C(=O)CH₂CH₃ 3) H₂O, H₃O⁺

   cyclohex-2-enone → 1) CH₃Li 2) H₂O, H₃O⁺

   chlorobenzene → Li

   bromobenzene → 1) Mg/ether 2) CH₃C(=O)CH₃ 3) H₂O, H₃O⁺

4. Give three different ways to prepare a tertiary alcohol.

5. Provide a mechanism for a Claisen condensation.

6. Draw the structure of the major organic product (if any) in each case:

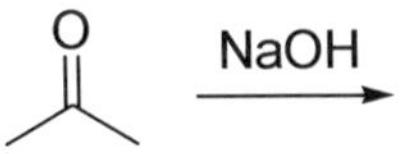

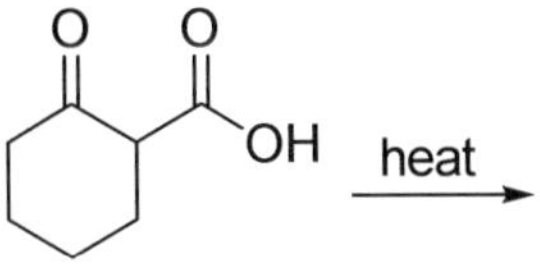

   acetone + NaOH

   2-oxocyclohexane-1-carboxylic acid (β-ketoacid) + heat

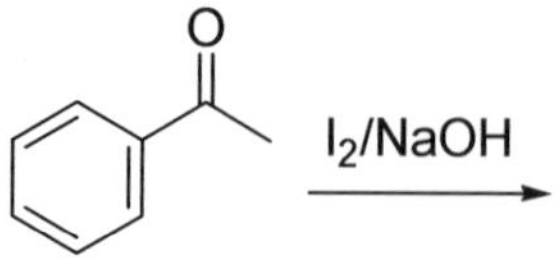

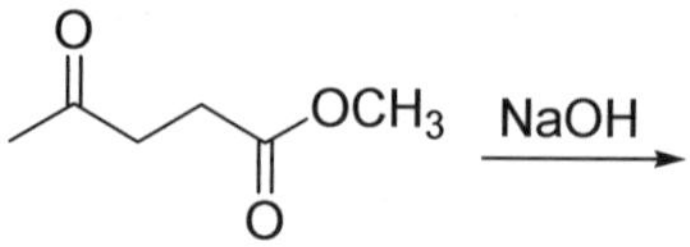

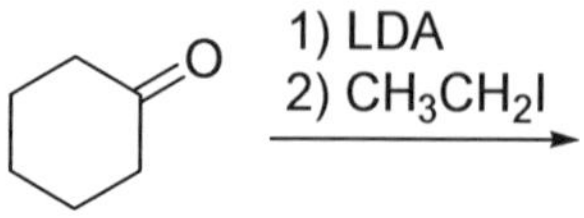

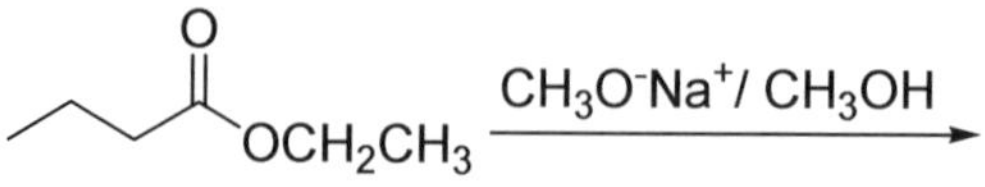

7. How would you prepare the following compounds:

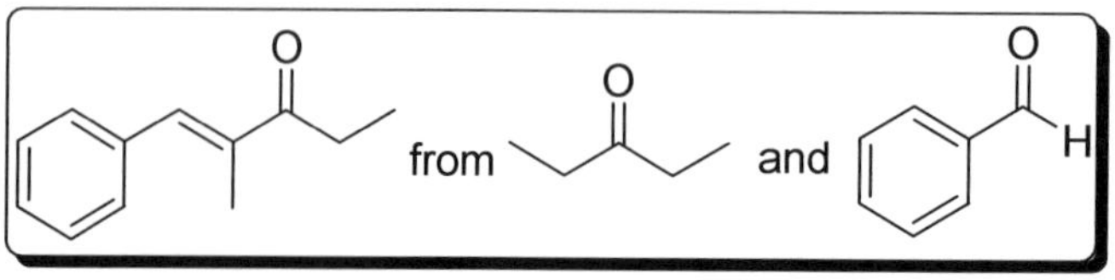

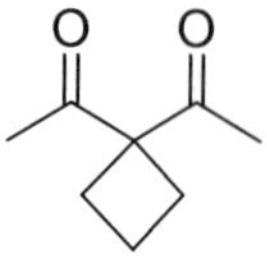

from: organic compounds containing five carbon atoms or less and inorganic reagents.
Hint: You will need a dihaloalkane as an electrophile.

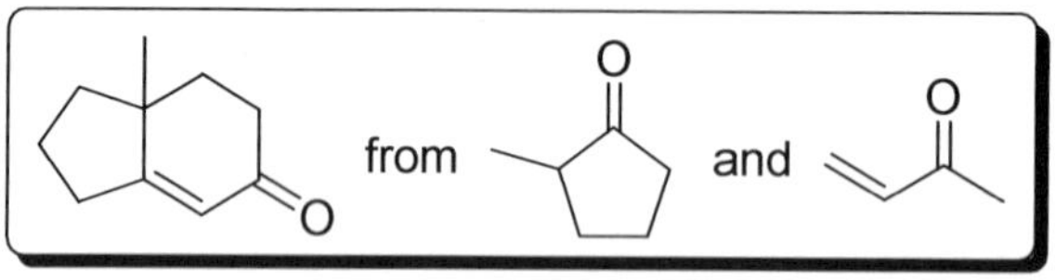

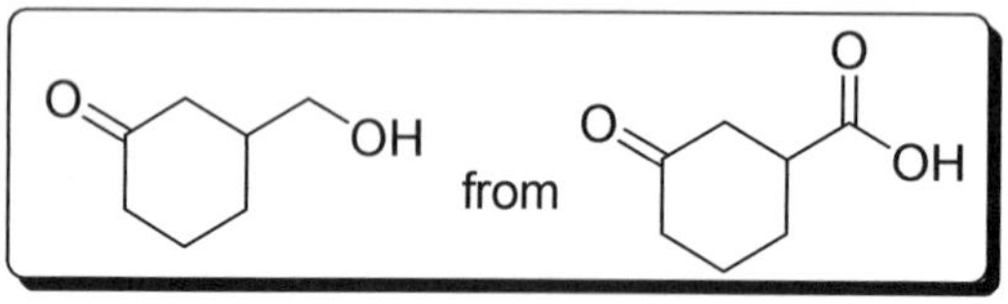

8. Why the following reaction cannot proceed in a good yield?

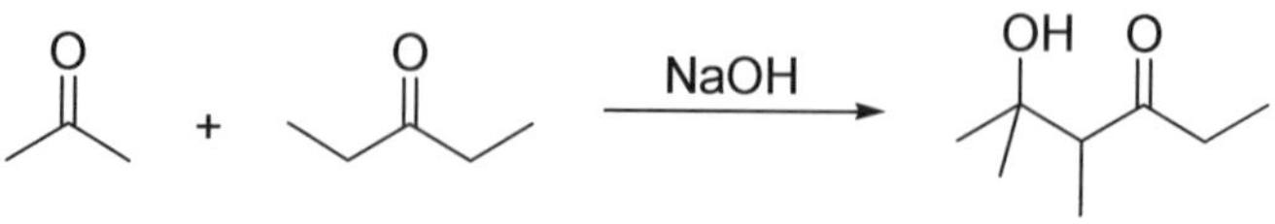

9. How would you prepare the compound below from organic compounds containing no more than four carbon atoms and inorganic reagents?

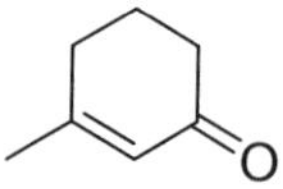

10. Draw the structures of missing compounds:

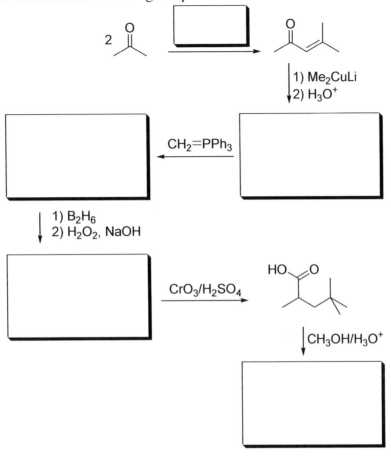

11. Why a carboxylic acid cannot be converted into the corresponding acyl chloride under ordinary conditions of nucleophilic acyl substitution?
12. Provide a mechanism for conversion of an acyl chloride into an ester.
13. Provide a mechanism for an acid-catalyzed ester hydrolysis.
14. Which monomers are used to make the following polymers?

15. Identify the smallest repeating unit in each of the polymers above.
16. Draw the structures of missing compounds:

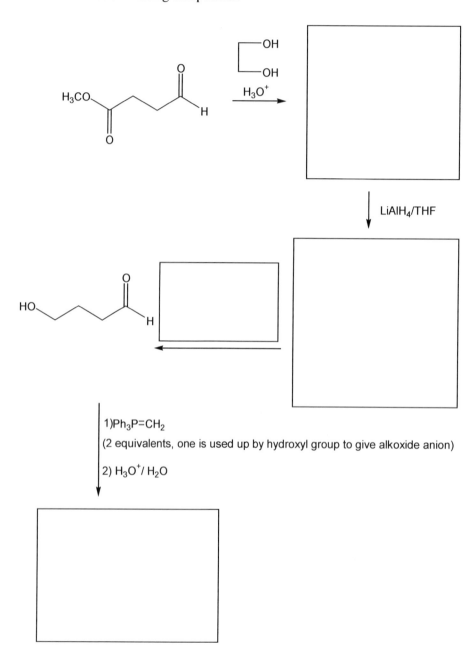

17. Complete the following table giving the structure or name for each molecule listed.

Name	Structural formula
	(structure: 4-hydroxypentanal — CH₃CH(OH)CH₂CH₂CHO)
acetophenone	
	(structure: N,N-dimethylpropanamide)
3-oxohexanoic acid	

Test 6 Solutions

1. A mechanism for NaBH₄ reduction of an aldehyde.

2. Preparation of methylenecyclopentane:

   cyclopentanone —Ph₃P=CH₂→ methylenecyclopentane

3. Reaction products:

   CH₃CH₂CH₂Br 1) Mg/ether 2) CH₃CH₂C(=O)CH₂CH₃ 3) H₂O, H₃O⁺ → tertiary alcohol

   2-cyclohexenone 1) CH₃Li 2) H₂O, H₃O⁺ → 3-methylcyclohexanone

   PhCl + Li → PhLi

   PhBr 1) Mg/ether 2) CH₃C(=O)CH₃ 3) H₂O, H₃O⁺ → 2-phenyl-2-propanol

Exercises in Organic Chemistry

4. Three different ways to prepare a tertiary alcohol:

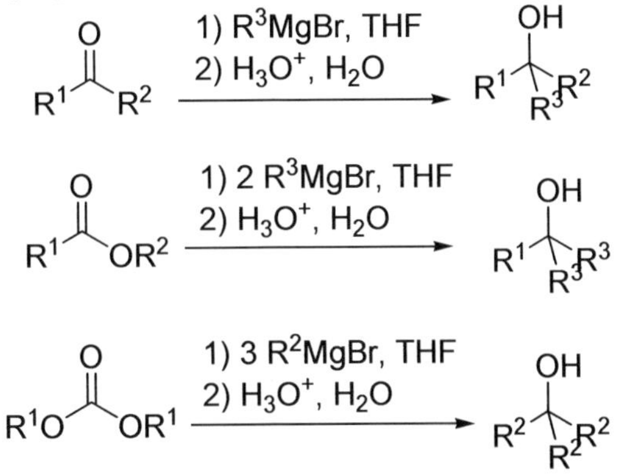

$$R^1-CO-R^2 \xrightarrow[\text{2) } H_3O^+, H_2O]{\text{1) } R^3MgBr, THF} R^1R^2R^3C-OH$$

$$R^1-CO-OR^2 \xrightarrow[\text{2) } H_3O^+, H_2O]{\text{1) 2 } R^3MgBr, THF} R^1(R^3)_2C-OH$$

$$R^1O-CO-OR^1 \xrightarrow[\text{2) } H_3O^+, H_2O]{\text{1) 3 } R^2MgBr, THF} (R^2)_3C-OH$$

5. Mechanism for a Claisen condensation:

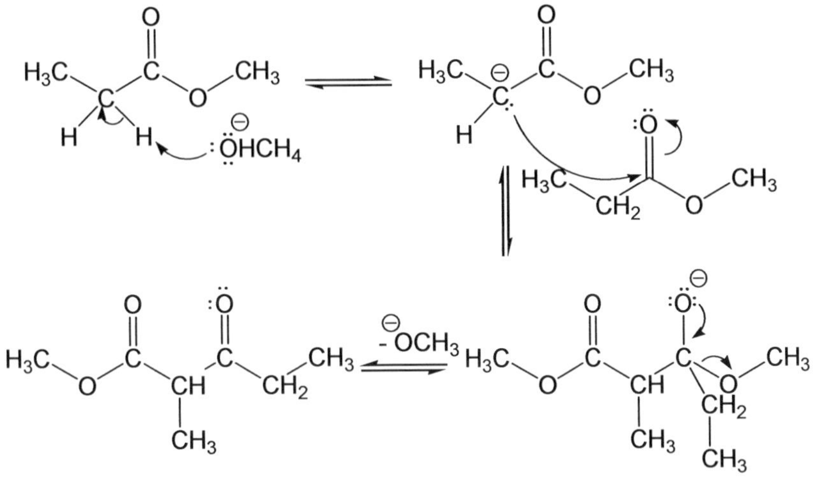

[mechanism scheme showing deprotonation of methyl acetate by methoxide, enolate attack on a second methyl acetate, tetrahedral intermediate, and loss of −OCH₃ to give methyl acetoacetate]

6. The structure of the major organic products:

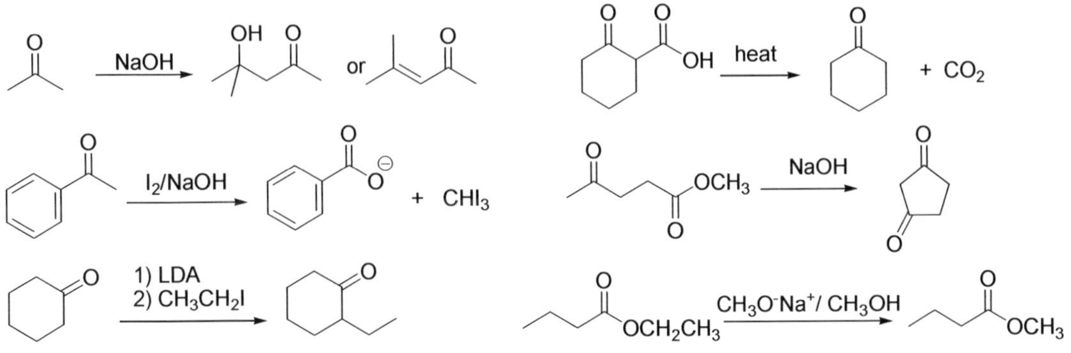

acetone + NaOH → 4-hydroxy-4-methylpentan-2-one or 4-methylpent-3-en-2-one

2-carboxycyclohexanone + OH, heat → cyclohexanone + CO₂

acetophenone + I₂/NaOH → benzoate anion + CHI₃

methyl 4-oxopentanoate + NaOH → 1,3-cyclopentanedione

cyclohexanone: 1) LDA, 2) CH₃CH₂I → 2-ethylcyclohexanone

ethyl butanoate + CH₃O⁻Na⁺ / CH₃OH → methyl butanoate

277

7. How would you prepare the following compounds:

8. Each of the two carbonyl compounds can generate the corresponding enolate ion. The two enolate ions can add to either of the two ketones. Therefore, four products will be obtained since each carbonyl can serve as either a nucleophile, or an electrophile.

9. Preparation of 3-methyl-2-cyclohexenone:

10. The structures of missing compounds:

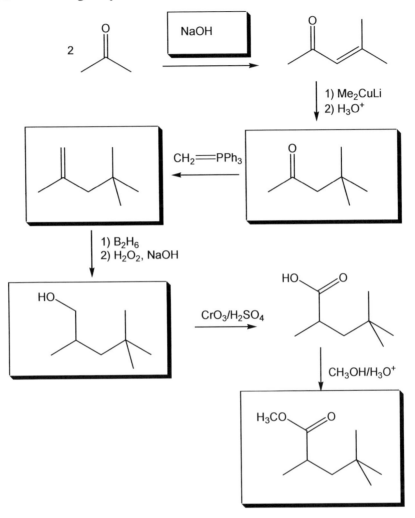

11. Carboxylic acid is more stable compared to the corresponding acyl chloride.

The tetrahedral intermediate will lose the best leaving group, which is Cl⁻, to give the starting materials. Chloride anion is a weak base and, therefore, a good leaving group, while the hydroxide anion is a strong base and, therefore, a poor leaving group.

12. A mechanism for conversion of an acyl chloride into an ester:

13. A mechanism for an acid-catalyzed ester hydrolysis:

14. Monomers:

15. Smallest repeating units:

16. The structures of missing compounds:

[Reaction scheme: methyl 4-oxobutanoate (H3CO-C(O)-CH2CH2-CHO) + HOCH2CH2OH / H3O+ → methyl ester with cyclic acetal (H3CO-C(O)-CH2CH2-CH(OCH2CH2O)); then LiAlH4/THF → HO-CH2CH2CH2-CH(OCH2CH2O); then H3O+, H2O → HO-CH2CH2CH2-CHO (4-hydroxybutanal); then 1) Ph3P=CH2 (2 equivalents, one is used up by hydroxyl group to give alkoxide anion), 2) H3O+/H2O → HO-CH2CH2CH2-CH=CH2]

17. Complete the following table giving the structure or name for each molecule listed.

Name	Structural formula
4-hydroxypentanal	CH3CH(OH)CH2CH2CHO
acetophenone	C6H5-C(O)-CH3
N,N-dimethylpropanamide	(CH3)2N-C(O)-CH2CH3
3-oxohexanoic acid	CH3CH2CH2-C(O)-CH2-COOH

Veljko Dragojlovic

Exercises in Organic Chemistry

24. Natural Products

Concepts to review: commonly occurring natural products, lipids, carbohydrates, amino acids, proteins.

Questions 6-15.

1. Amino acid aspargine has pKa values for –COOH and –NH$_3^+$ of 2.02 and 8.84, respectively. At what pH aspargine has the lowest solubility?

2. Partial hydrolysis with sodium hydroxide of a pentapeptide gave the following peptides:

    Val-Asp-Phe +   Phe-His +   Ala-Val

    What is the primary structure of the pentapeptide?

3. What is meant by the following terms as they apply to proteins?

    a) Secondary structure
    b) Tertiary structure
    c) Quaternary structure

4. The following compound is:

    a) an aldopentose.
    b) a ketopentose.
    c) an aldohexose.
    d) a ketopentose.

    ```
       CH₂OH
        |
        C=O
        |
    H---C---OH
        |
    H---C---OH
        |
       CH₂OH
    ```

5. Draw the following monosaccharide in its pyranose form (Haworth projection).

    ```
          O    H
           \\ //
            C
            |
    HO------C------H
            |
    HO------C------H
            |
    H-------C------OH
            |
    H-------C------OH
            |
           CH₂OH
    ```

6. Is the following compound a naturally occurring monosaccharide? Explain your answer.

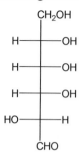

7. Humans and other mammals cannot utilize cellulose as food because it contains:

   a) α-1,4'–glucosidic bonds.
   b) β-1,4'–glucosidic bonds.
   c) α-1,6'–glucosidic bonds.
   d) β-1,6'–glucosidic bonds.

8. What class of natural products the following molecule belongs to? Describe its properties and biological activity.

9. Mark off the isoprene units in the following compounds. Classify the following compounds as monoterpenes, sesquiterpenes, diterpenes, etc.

10. Waxes are:

   a) triacylglycerides.
   b) amides.
   c) esters.
   d) ethers.

Answers.

1. An amino acid has the lowest solubility at its isoelectric point. Therefore, the actual question is: "What is pI of aspargine?" and the answer is:

$$pI = (pKa_1 + pKa_2)/2 = (2.02 + 8.84)/2 = 5.43$$

2. Primary structure of a peptide is its amino acid sequence. As was the case with the previous question, we have to "translate" is and ask the actual question ("What is the amino acid sequence of the pentapeptide?"). The answer is:

	Val–	Asp–	Phe	
			Phe–	His
Ala–	Val			

primary structure: Ala – Val – Asp – Phe – His

3. a) Secondary structure of a peptide is conformation (spatial arrangement) of segments of protein chain. Examples of secondary structure include α-helix, β-pleated sheet and random coil. b) Tertiary structure of a protein is its overall shape. Based on their tertiary structure, proteins are divided into globular and fibrous. c) Some proteins are composed of more than one unit (chain). Such assemblies have quaternary structure. Quaternary structure indicates both the number of units (dimer, trimer, tetramer and so on) as well as their arrangement in space (planar, tetrahedral and the like).

4. b) It has a five carbon atoms. Therefore, it is an pentose. Keto group means that it is also a ketose. Therefore, overall it is a ketopentose.

5. In a Haworth projection, rings is represented as being planar and we use dash bonds to indicate perspective. First, we need to identify the pyranose form. It is a six-member ring that conatins five carbon and one oxygen atoms. It includes the aldehyde carbon, four other carbon atoms and an oxygen atom:
    Next, we should make a hemicaetal bond between the oxygen labeled as atom 6 and the carbonyl carbon and rotate the molecule 90° clockwise:

Finally, we should convert the drawing into Haworth conformation by drawing a regular hexagon and adding wedge bonds:
    To make sure that you get the correct stereochemistry, you may want to indicate stereochemistry of each bond (recall that in the original Fischer projection horizontal bonds project toward the viewer and terminal vertical bonds away from the viewer. Internal vertical bonds are in the plane of the drawing). Finally, keep in mind that you have to rotate C4-C5 bond to get Newman

(planar) projection, which places CH$_2$OH above and H below the plane of the ring. Hydroxyl group on C1 (anomeric carbon) can be either above or below the plane of the ring. Here it is shown in a more stable configuration.

6. A difference between the natural and non-natural monosaccharides is that natural have D configuration, while non-natural have L. In a Fischer projection, monosaccharide with a D configuration has –OH substituent on the right on the highest numbered chirality center, while monosaccharide with L configuration has it on the left. At first glance it may appear that this is a non-natural monosaccharide. However, this is not a conventional Fischer projection! In a Fischer projection of monosaccharide, the aldehyde group should be on the top and -CH$_2$OH on the bottom. This molecule is represented up side down (allowed transformation of a Fischer projection). To get the conventional representation, we need to rotate it 180°. Now we can see that –OH group on the carbon 5 (highest numbered chirality center) is on the right and that this is a natural carbohydrate.

7. b).

8. This is lipid, or more specifically a prostaglandin. Prostaglandins are biologically active compounds. They are active in a very low concentration and are present in all cells. Prostaglandins play a role in regulation of blood pressure, inflammatory processes and contractions of uterus during childbirth.

9. Isoprene units and classification:

sesquiterpene

monoterpene

diterpene

10. c).

Exercises in Organic Chemistry

Multiple Choice Practice Test: Chapters 1-22.

Directions
• Exam is 2 hours long. • Do not use calculator, periodic table, table of chemical shifts or any other aid. • Make no marks in the book. Use scrap paper.

1. Which of the following carbocations is most likely to undergo a rearrangement?

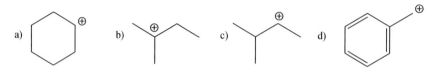

2. What is the IUPAC name of the following compound?

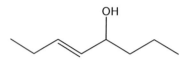

   a) (E)-5-octene-4-ol
   b) (E)-3-octene-5-ol
   c) (Z)-5-octene-4-ol
   d) (Z)-3-octene-5-ol

3. Which compound is the most resistant to oxidation by KMnO$_4$?

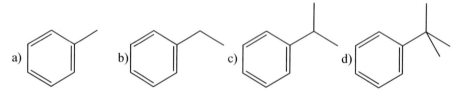

4. The following carbon-carbon bond is formed by overlap of:

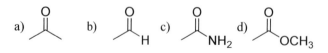

   a) sp^3-sp^2 orbitals.
   b) sp^3-sp orbitals.
   c) p-sp^2 orbitals.
   d) sp-sp^2 orbitals.

5. Which of the following compounds has the highest boiling point?

6. Which movement of electrons represents a reasonable reaction step?

7. Which of the following is the most stable form of 1-*t*-butyl-4-methylcyclohexane?

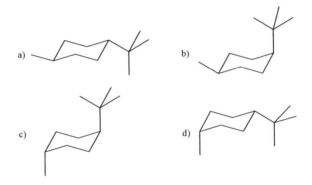

8. How many stereoisomers of this compound are possible?

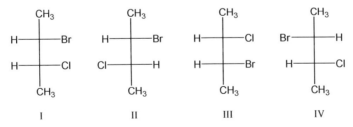

a) 2
b) 4
c) 6
d) 8

9. Which of the following compounds are enantiomers?

	CH₃			CH₃			CH₃			CH₃	
H—		—Br	H—		—Br	H—		—Cl	Br—		—H
H—		—Cl	Cl—		—H	H—		—Br	H—		—Cl
	CH₃			CH₃			CH₃			CH₃	
	I			II			III			IV	

a) I and II
b) I and IV
c) II and III
d) II and IV

10. What is the relative configuration of the following molecule?

    a) 1R,2S
    b) 1S,2R
    c) 1R,2R
    d) 1S,2S

11. Which ion is an intermediate in conversion of cyclohexanol into chlorocyclohexane by S_N1 reaction mechanism?

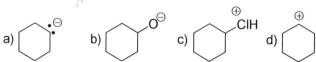

12. Which of the following ions is aromatic?

13. Which of the following is the correct transition state for an S_N2 reaction between bromomethane and sodium hydroxide? Charges, if any, are not shown.

14. Which of the following correctly represents electron flow for the bromination of an alkene?

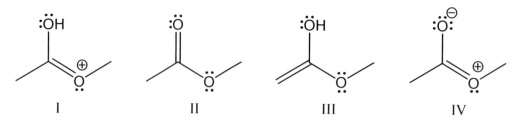

15. Which of the following represent resonance structures?

I   II   III   IV

    a) I and II
    b) I and IV
    c) II and III
    d) II and IV

16. Which of the following is not a reasonable electron flow?

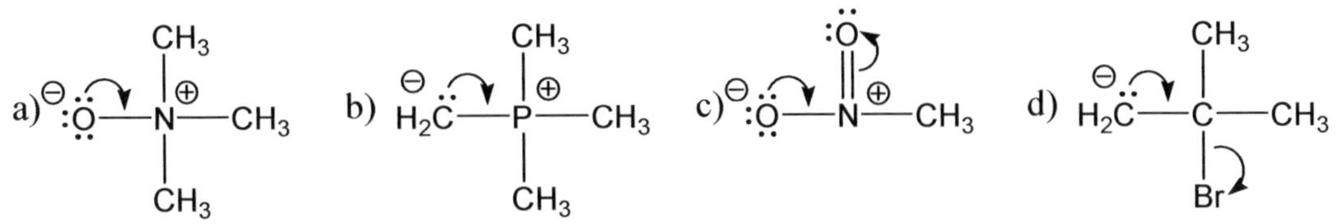

a) ⊖:Ö⤴N⁺(CH₃)(CH₃)—CH₃   b) ⊖:H₂C⤴P⁺(CH₃)(CH₃)—CH₃   c) ⊖:Ö⤴N⁺(=O)—CH₃   d) ⊖:H₂C⤴C(CH₃)(Br)—CH₃

17. Neurontin (gabapentin), shown below, is an anticonvulsive medication manufactured by Pfizer. This compound is an:

   a) α-amino acid
   b) β-amino acid
   c) γ-amino acid
   d) δ-amino acid

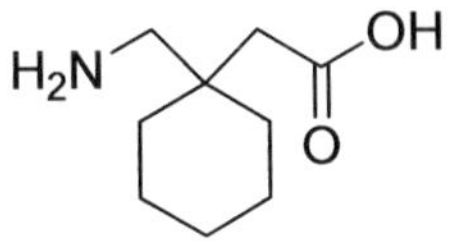

18. What is the major product of the following reaction?

1) CH₃ONa, CH₃OH
2) H₃O⁺, H₂O

a) CH₃CH(OCH₃)CH₂OH   b) CH₃CH(OH)CH₂OCH₃   c) CH₃CH(OCH₃)CH₂OCH₃   d) CH₃CH(OH)CH₂OH

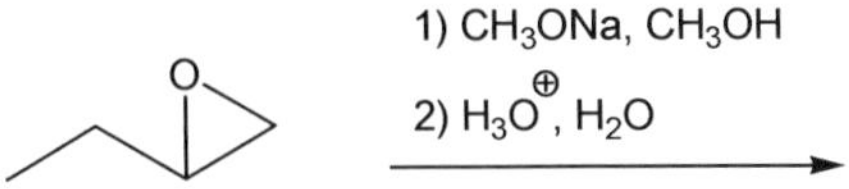

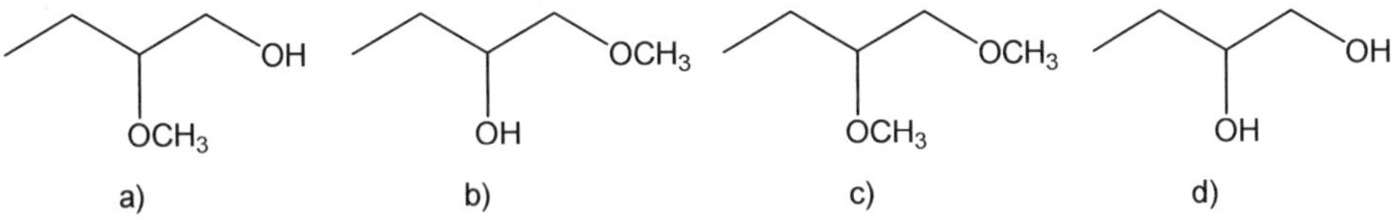

19. What is the major product of the following reaction?

H₃CHC=CHCH₃ →[CF₃CO₃H]

a) H₃CHC(OH)—CHCH₃(OH)   b) H₃CHC(OH)—CHCH₃(OCOCF₃)   c) H₃CC(=O)—C(=O)CH₃   d) H₃CHC—CHCH₃ (epoxide)

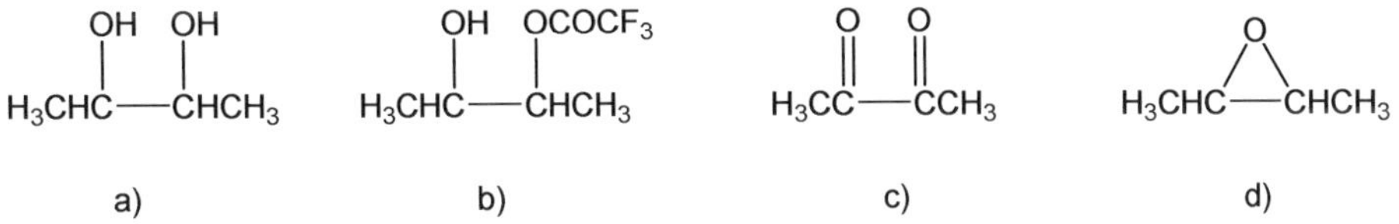

20. What is a suitable reagent to accomplish the following reaction?

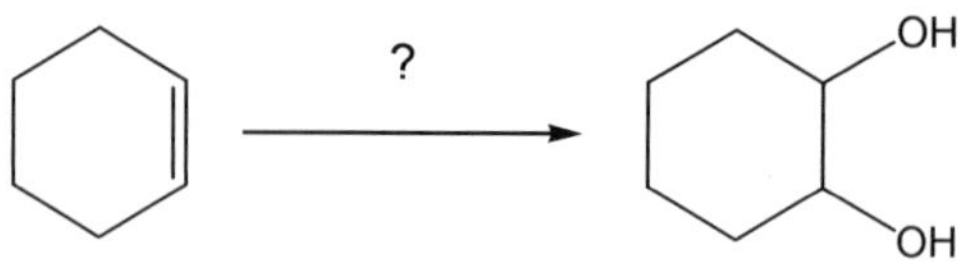

   a) HIO$_4$
   b) OsO$_4$
   c) O$_3$, CH$_3$OH followed by NaBH$_4$, CH$_3$OH
   d) CrO$_3$, H$_2$SO$_4$

21. What is the major product of the following reaction?

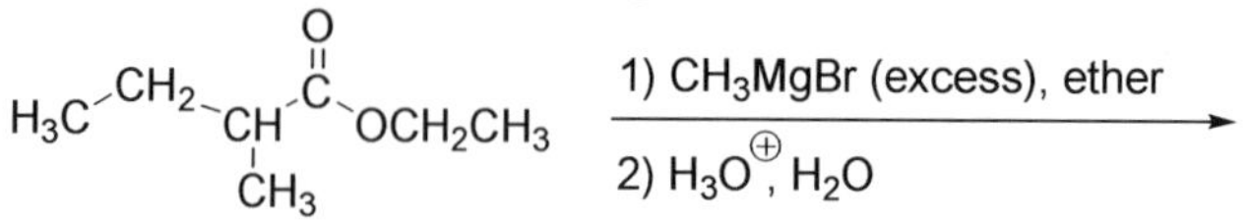

   H$_3$C-CH$_2$-CH(CH$_3$)-C(=O)-OCH$_2$CH$_3$ → 1) CH$_3$MgBr (excess), ether; 2) H$_3$O$^+$, H$_2$O

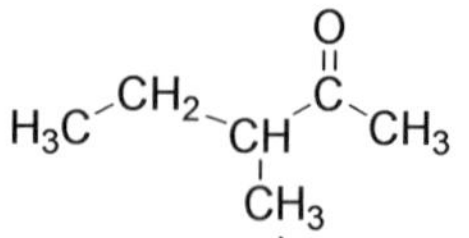

   a) H$_3$C-CH$_2$-CH(CH$_3$)-C(=O)-CH$_3$
   b) H$_3$C-CH$_2$-CH(CH$_3$)-C(OH)(OCH$_3$)(OCH$_3$)
   c) H$_3$C-CH$_2$-CH(CH$_3$)-C(OH)(CH$_3$)(CH$_3$)
   d) H$_3$C-CH$_2$-CH(CH$_3$)-C(OH)(CH$_3$)(OCH$_2$CH$_3$)

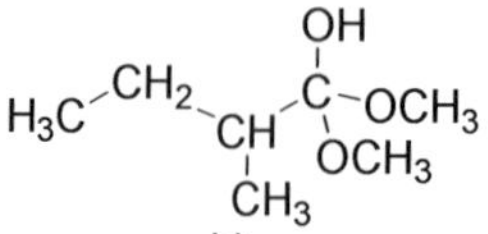

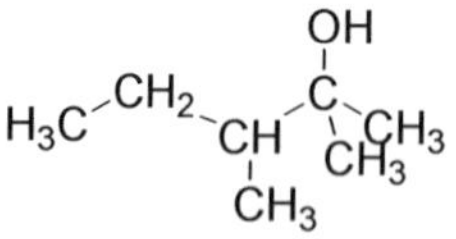

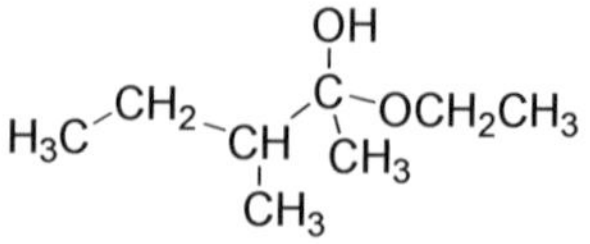

22. What is a suitable starting material for the following reaction?

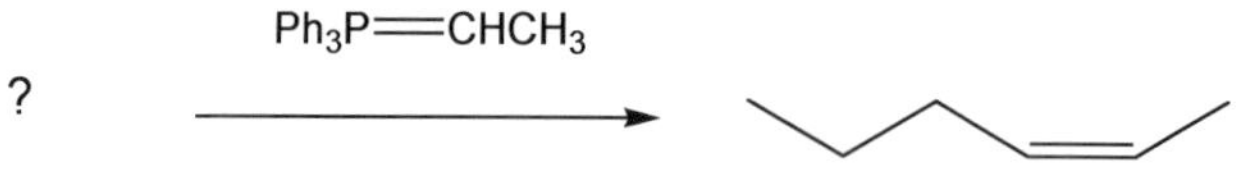

   ? $\xrightarrow{Ph_3P=CHCH_3}$

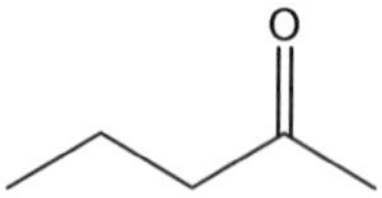

   a) 2-pentanone
   b) methyl butanoate (with OCH$_3$)
   c) butanal (with H)
   d) butanoic acid (with OH)

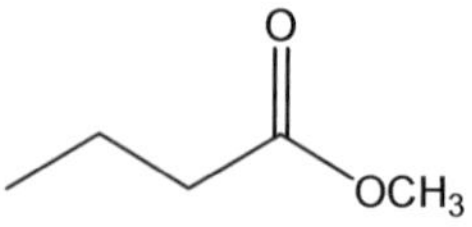

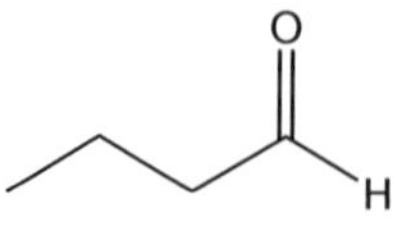

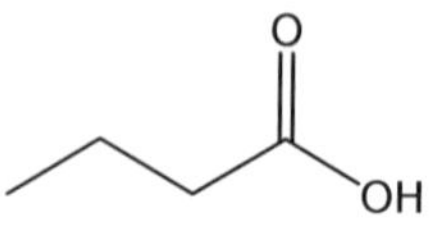

23. The best reagent to accomplish the following reaction is

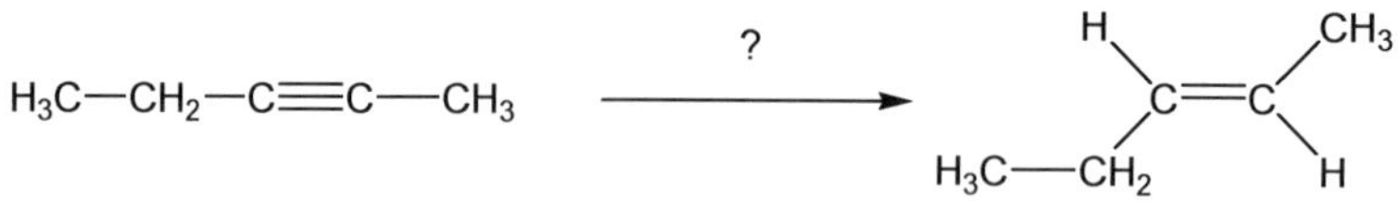

   H$_3$C—CH$_2$—C≡C—CH$_3$ → cis-alkene (H$_3$C—CH$_2$ and CH$_3$ on same side)

   a) Na, NH$_3$
   b) NaNH$_2$
   c) H$_2$, Pd/C, pyridine
   d) NaBH$_4$, CH$_3$CH$_2$OH

24. Which of the following substituents activates benzene ring towards electrophilic aromatic substitution?

   a) $-NH_3^+$
   b) $-NH_2$
   c) $-NO_2$
   d) $-N(CH_3)_3^+$

25. Which reaction product will contain a six-membered ring?

   a) 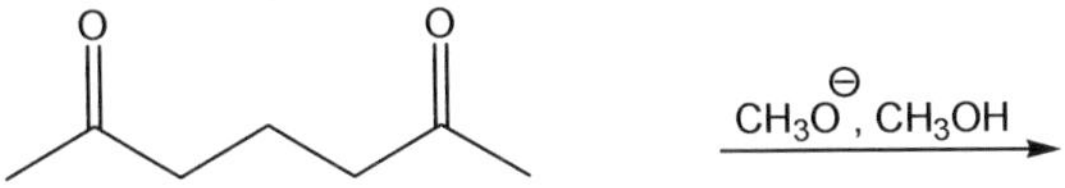 [diketone structure] $\xrightarrow{CH_3O^\ominus, CH_3OH}$

   b) 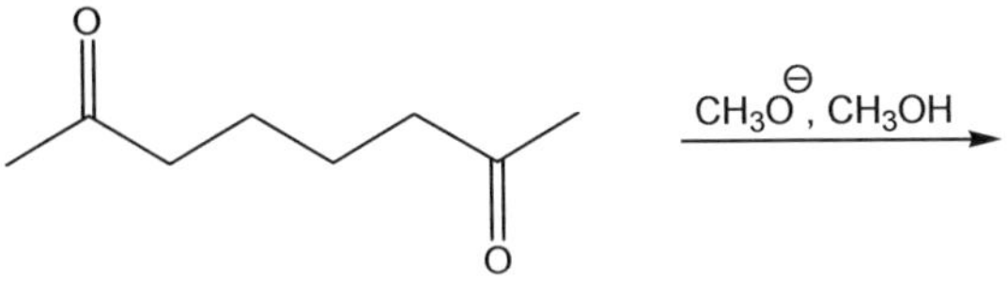 [diketone structure] $\xrightarrow{CH_3O^\ominus, CH_3OH}$

   c) 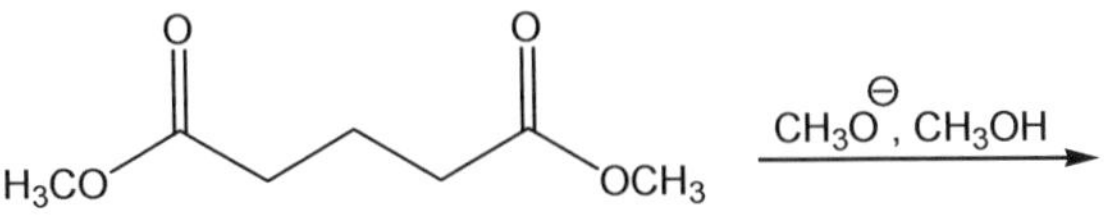 H_3CO—[diester structure]—OCH_3 $\xrightarrow{CH_3O^\ominus, CH_3OH}$

   d) 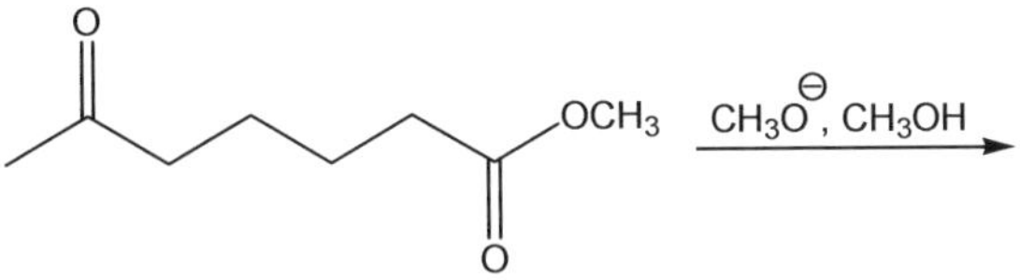 [ketoester structure]—OCH_3 $\xrightarrow{CH_3O^\ominus, CH_3OH}$

26. Which of the following reactions involves free radical intermediates?

   a) $CH_3CH_2CH=CH_2 + HBr \rightarrow CH_3CH_2CHBrCH_3$
   b) $CH_3CH_2CH=CH_2 + HBr \rightarrow CH_3CH_2CH_2CH_2Br$
   c) $CH_3CH_2CH=CH_2 + HBr + CH_3OH \rightarrow CH_3CH_2CH(OCH_3)CH_2Br$
   d) $CH_3CH_2CH=CH_2 + HBr + CH_3OH \rightarrow CH_3CH_2CHBrCH_2OCH_3$

27. What is the best reaction sequence to prepare 1-bromo-3-nitrobenzene starting with benzene?

   a) 1) HBr, FeBr$_3$; 2) HNO$_3$, H$_2$SO$_4$
   b) 1) Br$_2$, FeBr$_3$; 2) HNO$_3$, H$_2$SO$_4$
   c) 1) HNO$_3$, H$_2$SO$_4$; 2) HBr, FeBr$_3$
   d) 1) HNO$_3$, H$_2$SO$_4$; 2) Br$_2$, FeBr$_3$

28. Which structure is not an intermediate in nitration of phenol?

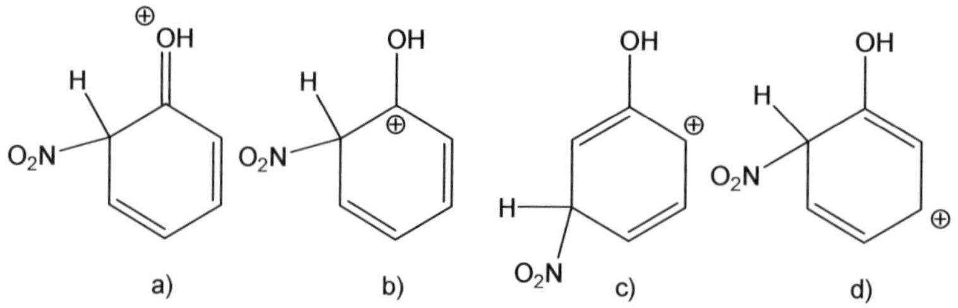

a) b) c) d)

29. What is the major product of the following reaction?

1) $(CH_3)_3CuLi$, THF
2) $H_3O^{\oplus}$, H_2O

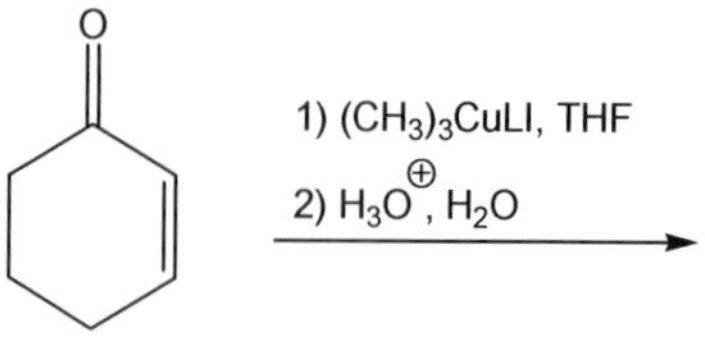

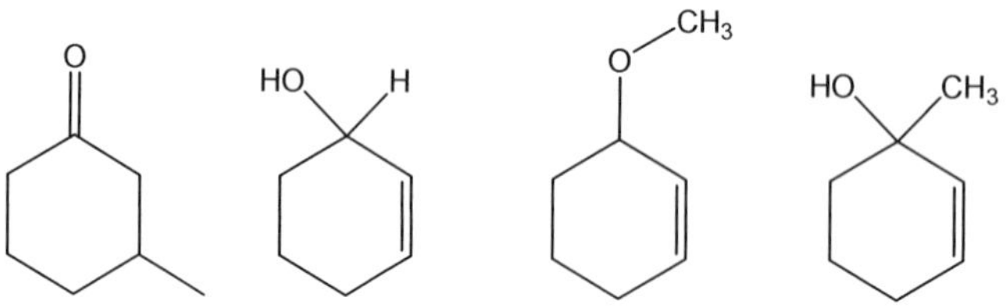

a) b) c) d)

30. Which of the following reactions is a suitable method for preparation of an acid anhydride?

a) $CH_3COOH + CH_3COO^-$
b) $CH_3COCl + CH_3COO^-$
c) $CH_3COOH + CH_3CONH_2$
d) $CH_3COOH + CH_3COOCH_3$

31. Which of the following carbonyl compounds reacts the fastest with nucleophiles?

a) C_6H_5CHO
b) $HCHO$
c) CH_3COCH_3
d) $C_6H_5COC_6H_5$

32. What is the major product of the following reaction?

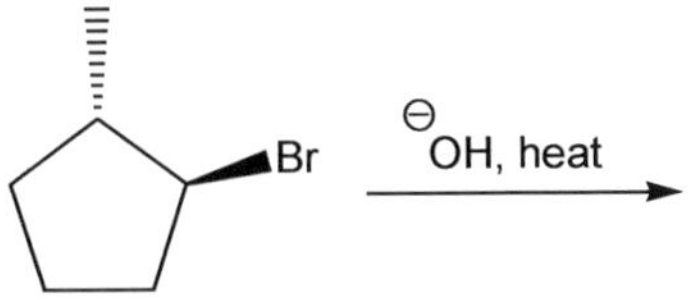

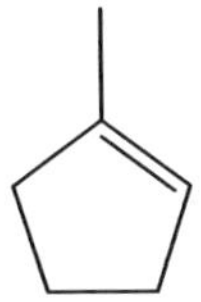

[cyclopentane with Br (wedge up) + ⁻OH, heat]

a) [1-methylcyclopentene-like: cyclopentene with substituent] b) [cyclopentene with substituent] c) [cyclopentene] d) [cyclopentanol with OH]

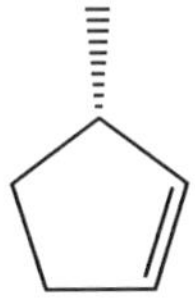

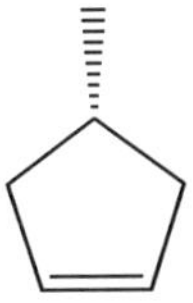

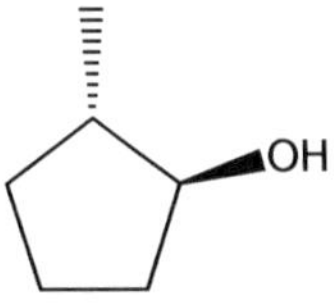

33. What is a suitable reagent for the following reaction?

[tetrahydrothiophene] → ? → [S-methyl sulfonium with I⁻]

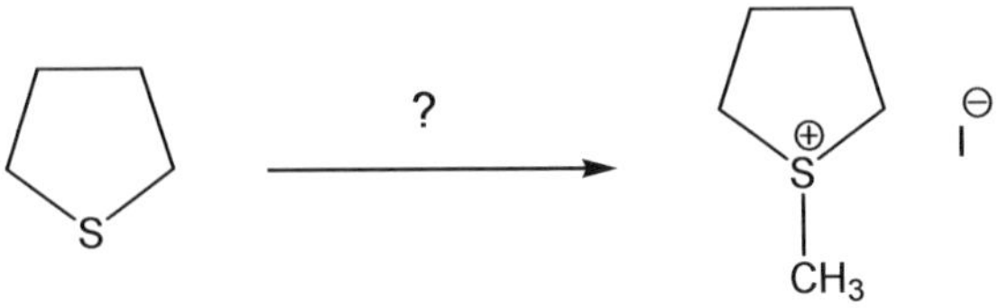

a) CH_3OH, HI
b) CH_3ONa, NaI
c) CH_3F, NaI, acetone
d) CH_3I

34. Which of the following alcohols cannot be oxidized with Jones reagent (CrO_3 in concentrated H_2SO_4)?

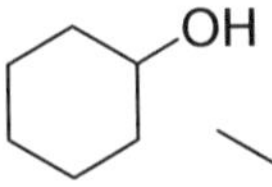

a) cyclohexanol   b) neopentyl-type diol/alcohol   c) benzyl alcohol   d) 1-methylcyclopentanol

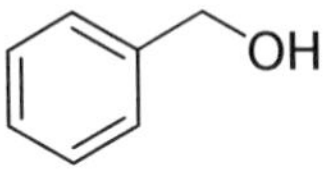

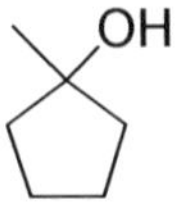

35. What is the best reagent for the following reaction?

[nitrobenzene (NO_2)] → ? → [aniline (NH_2)]

a) $KMnO_4$, NaOH, 0°C
b) H_2SO_4
c) Zn, HCl followed by NaOH
d) $NaNO_2$, HCl, 0°C

295

36. What is the major product of the following reaction?

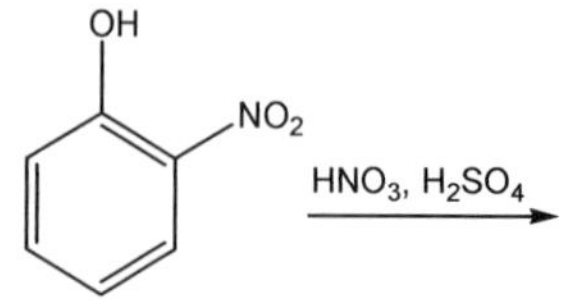

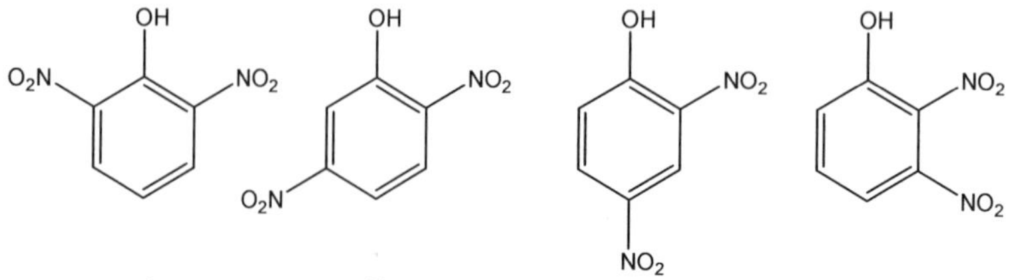

37. What is the product of the following sequence of reactions?

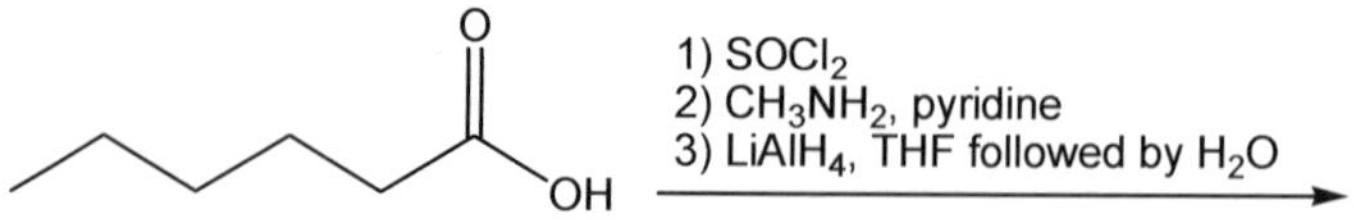

Reagents: 1) SOCl$_2$; 2) CH$_3$NH$_2$, pyridine; 3) LiAlH$_4$, THF followed by H$_2$O

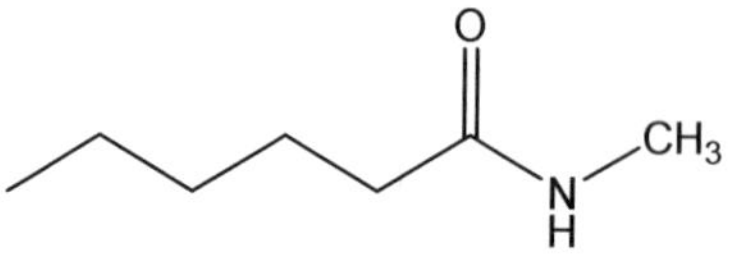

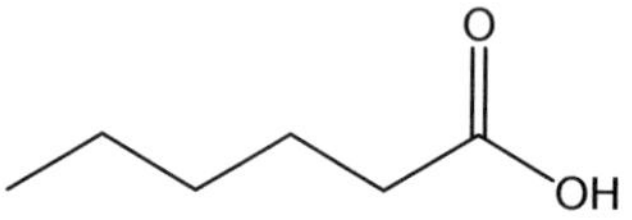

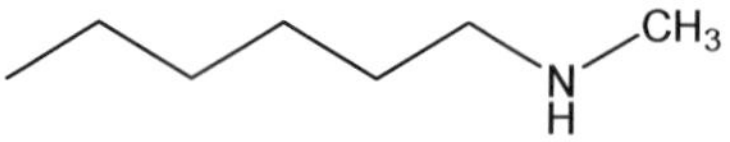

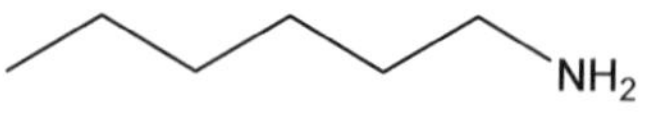

38. Which reagent is suitable for the following reaction?

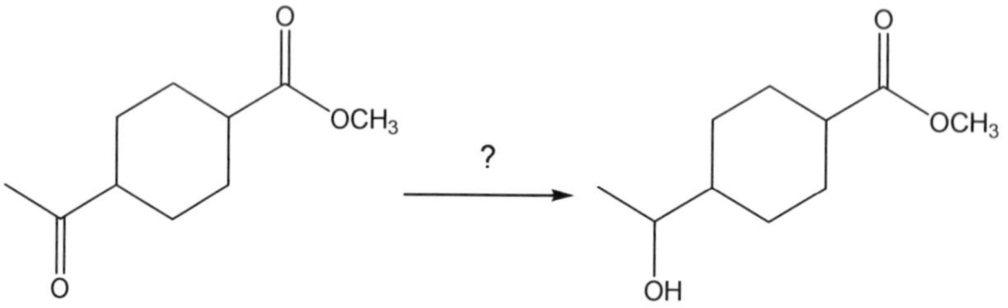

a) NaBH$_4$, CH$_3$OH
b) LiAlH$_4$, THF followed by H$_3$O$^+$, H$_2$O
c) diisobutylaluminum hydride (DIBAL-H), THF, -78°C
d) NaBH$_3$CN

39. Which compound does not react with sodium?

   a) CH_3-O-CH_3
   b) CH_3CH_2OH
   c) CH_3COOH
   d) $CH_3C\equiv CCH_3$

40. Which one of the following compounds is the most reactive towards an S_N2 reaction?

   a) $CH_3CH_2CH_2CH_2Cl$
   b) $(CH_3)_3CCl$
   c) $CH_3CHClCH_2CH_3$
   d) $(CH_3)_2CHCH_2Cl$

41. What is the product of the following reaction?

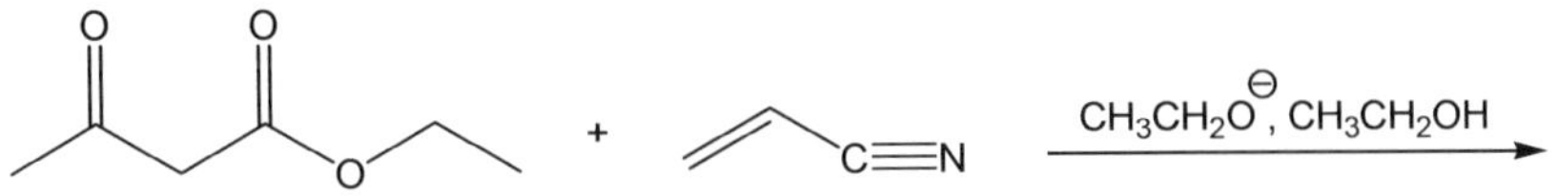

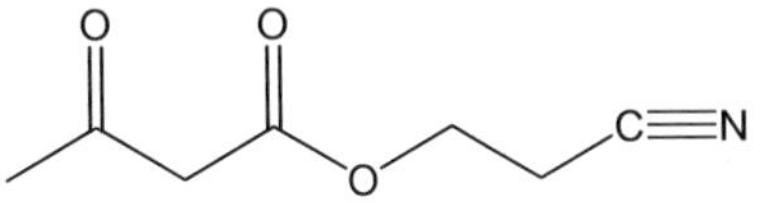

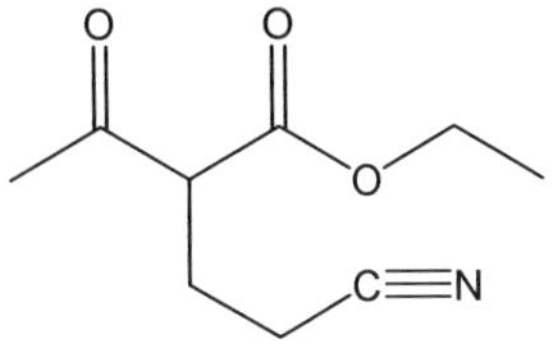

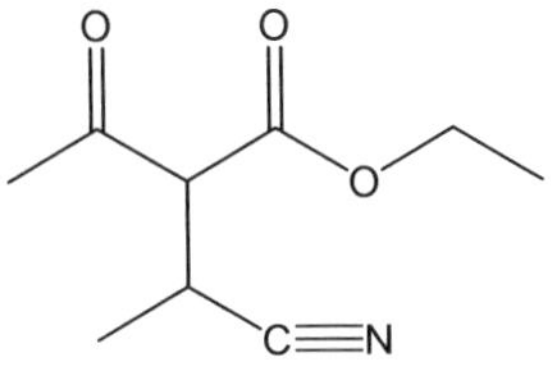

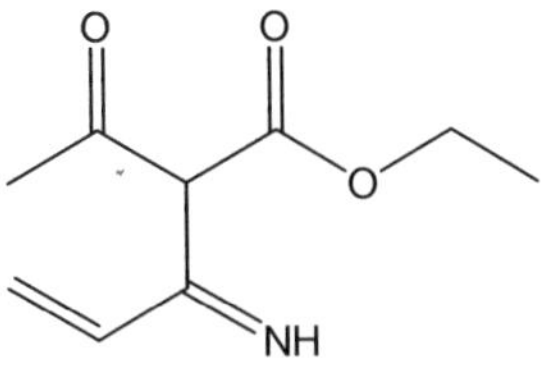

42. Which polymer is produced by the following reaction?

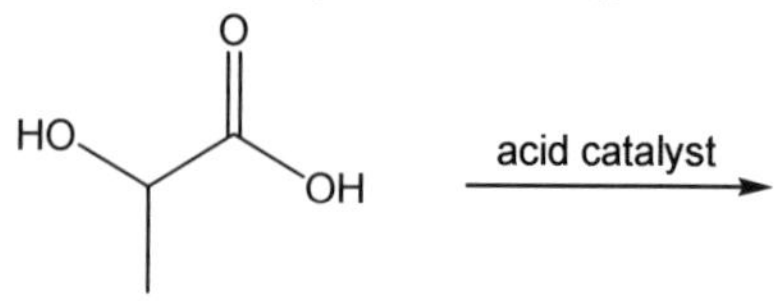

a)

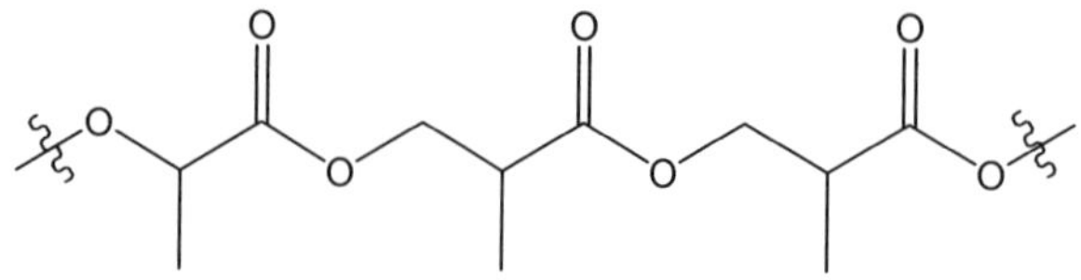

b)

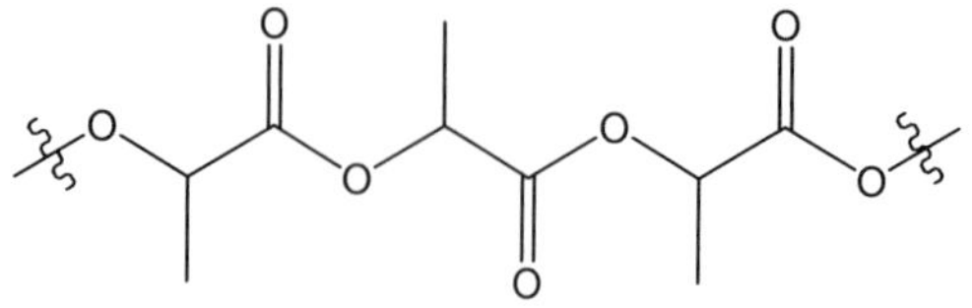

c)

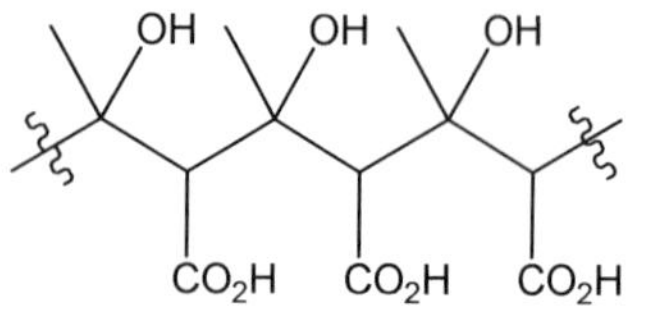

d)

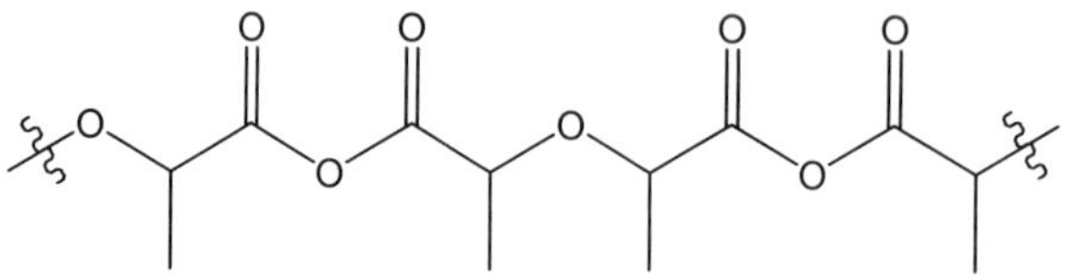

43. What is the major product of the following reaction?

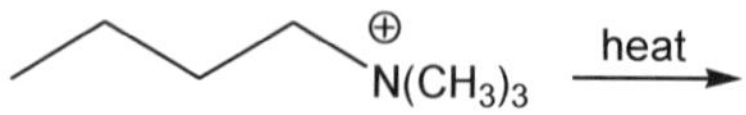

a)

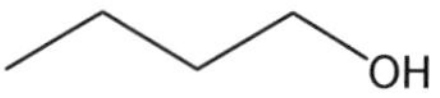

b)

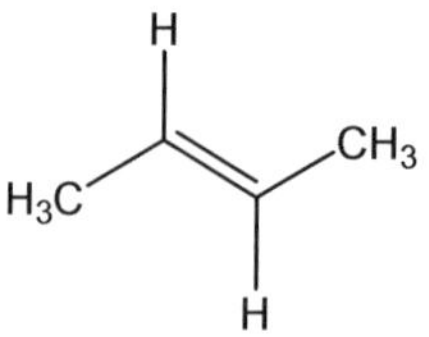

c)

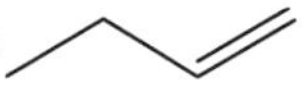

d)

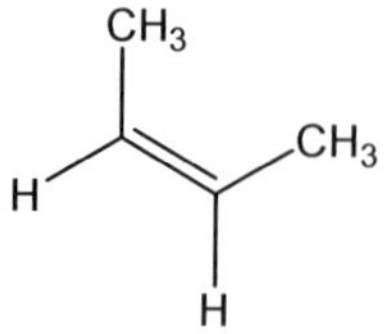

44. Which structure is consistent with the following ^{1}H and ^{13}C NMR spectra?

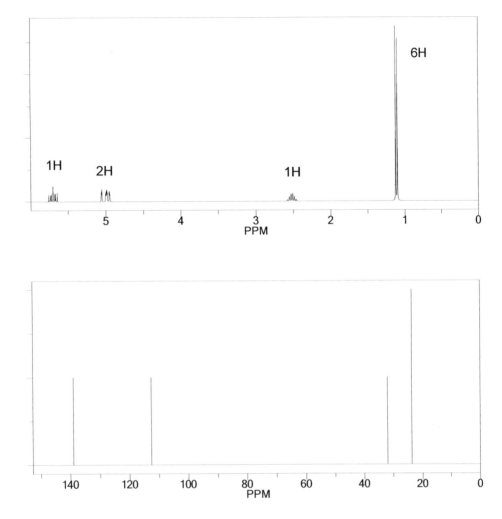

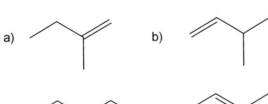

45. Which of the following is a suitable method for preparation of *t*-butyl ethyl ether?

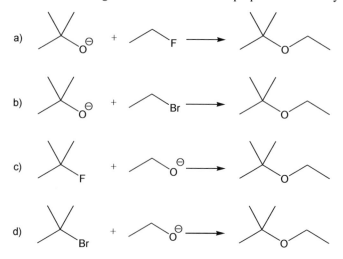

46. What are the relative rates of reaction of the following compounds with hydroxide ion in a polar aprotic solvent?

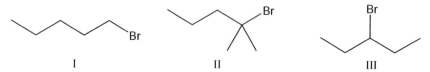

a) III<I<II
b) II<I<III
c) II<III<I
d) I<III<II

47. Which one is an incorrect completion of the following statement. In an S_N2 reaction between (*R*)-3-bromohexane and sodium hydroxide,

a) the rate of the reaction is independent of the concentration of sodium hydroxide.
b) the product is (*S*)-3-hexanol.
c) if (*R*)-3-bromohexane is replaced by (*S*)-3-chlorohexane the reaction rate decreases.
d) if (*R*)-3-bromohexane is replaced by (*R*)-3-methyl-3-bromohexane the reaction rate decreases.

48. Which of the following reactions is carried out in the presence of a heterogeneous catalyst?

a) Hydration of an alkyne
b) Hydrolysis of an alkyl halide
c) Hydrogenation of an alkene
d) Hydroxylation of an alkene

49. Which of the following compounds is not a nucleophile?

   a) CH$_3$OH
   b) H$_2$
   c) CN$^-$
   d) NH$_3$

50. Which structure is consistent with the following infrared and ^{13}C NMR spectra?

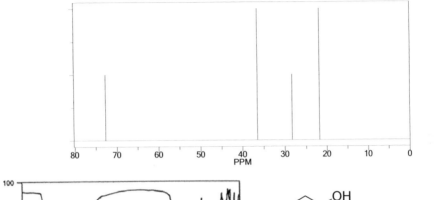
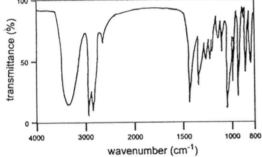

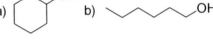
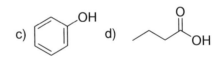

51. Which one of the following substituents has *meta*-directing effect in electrophilic aromatic substitution reactions?

   a) -OCH$_3$
   b) -OCOCH$_3$
   c) -COCH$_3$
   d) -N(CH$_3$)$_2$

52. Which compound is an amide?

   a) CH$_3$C(S)OCH$_3$
   b) CH$_3$CH$_2$CONH$_2$
   c) CH$_3$NH$_2$
   d) CH$_3$CN

53. Which structure is consistent with the following ¹H NMR spectrum?

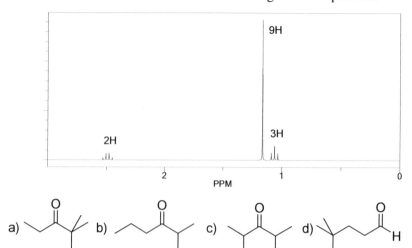

54. An example of a nucleophilic substitution reaction is:

   a) Reaction of phenol with bromine.
   b) Reaction of bromoethane with aqueous sodium hydroxide.
   c) Reaction of ethane with bromine
   d) Reaction an aldehyde with chromium trioxide in sulfuric acid.

55. What are the products of the following reaction?

56. Which structure is consistent with the following IR spectrum?

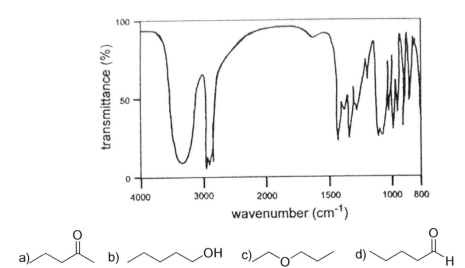

57. Which of the following is a termination step of a free radical reaction?

a) Cl· + CH₄ → CH₃· + HCl
b) CH₃· + HCl → CH₄ + Cl·
c) CH₃· + CH₃· → CH₃-CH₃
d) CH₃· + CH₄ → CH₄ + CH₃·

58. Which of the following compounds undergoes S$_N$1 reactions the fastest?

a) CH₃Cl
b) CH₂=CHCl
c) CH₃CH₂CH₂Cl
d) CH₂=CHCH₂Cl

59. The order of reactivity of the following compounds towards electrophilic aromatic substitution is:

I: C₆H₅Cl II: C₆H₅OH III: C₆H₅CH₃ IV: C₆H₅NO₂

a) IV < I < III < II
b) IV < III < I < II
c) II < III < I < IV
d) II < I < III < IV

60. Which of the following compounds gives a positive iodoform test?

   a) PhCH$_2$COCH$_3$
   b) PhCH$_2$COCH$_2$CH$_3$
   c) PhCH$_2$CH$_2$CHO
   d) PhCH$_2$CH$_2$OH

61. Which of the following compounds undergoes decarboxylation upon heating?

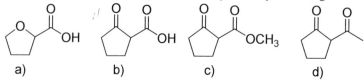

   a)   b)   c)   d)

62. Ozonization of cyclohexane followed by treatment with sodium peroxide gives:

   a) a diacid.
   b) a dialdehyde
   c) a diol
   d) a diketone

63. Nitronium ion, NO$_2^+$, is:

   a) an electrophile.
   b) a nucleophile.
   c) a Lewis base.
   d) a free radical.

64. Product of reaction of methyl 3-oxopentanoate with sodium borohydride in methanol has:

   a) two hydroxyl groups.
   b) two carbonyl groups.
   c) a carbonyl and a hydroxyl group.
   d) an ester and a hydroxyl group.

65. Which of the following alcohols can be oxidized by Jones reagent (CrO$_3$/H$_2$SO$_4$) to a ketone?

   a) 2-methyl-2-pentanol.
   b) 2-methyl-3-pentanol.
   c) 2-methyl-1-pentanol
   d) 3-methyl-1-pentanol

66. Which of the following substituents has an electron-donating resonance effect?

   a) –COCH$_3$
   b) –C(CH$_3$)$_3$
   c) –CN
   d) –OCH$_3$

67. Which of the following is the correct structure for 4-hydroxy-3-chloro-2-hexenal?

   a) $CH_3CH_2C=CClCHOHCH_2CHO$
   b) $CH_3CHOHCH=CClCH_2CHO$
   c) $CH_3CHOHCH_2CCl=CHCHO$
   d) $CH_3CH_2CHOHCCl=CHCHO$

68. In molecule of an amide, C-O-N bond angle is closest to:

   a) 90°
   b) 109°
   c) 120°
   d) 180°

69. What is the product of the following reaction sequence?

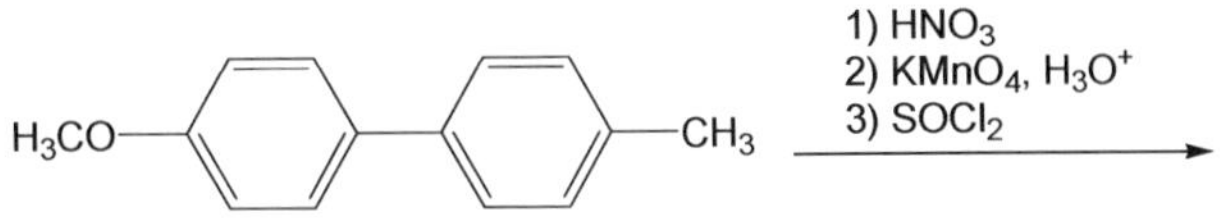

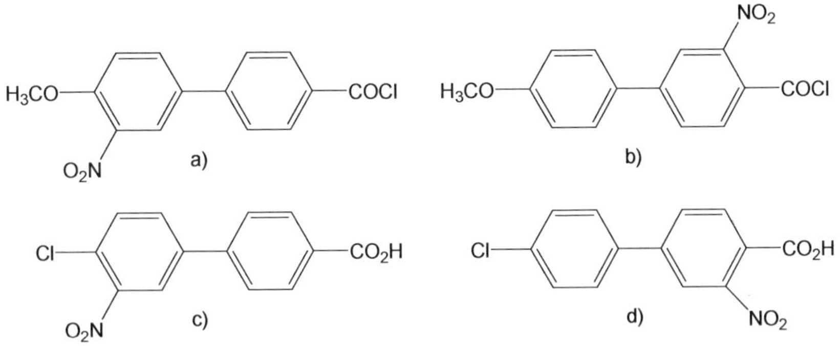

70. A compound A reacted with hydrogen and a palladium catalyst in presence of pyridine to give compound B. Compound B reacted with bromine to give 2,3-dibromoheptane. Compound A is:

   a) 1-heptene
   b) 2-heptene
   c) 1-heptyne
   d) 2-heptyne

End of the test. Answers are on the next page.

Answers to Multiple Choice Practice Test: Chapters 1-22.

1. c); 2. a); 3. d); 4. d); 5. c); 6. b); 7. a); 8. b); 9. d); 10. a); 11. d); 12. a); 13. a); 14. b); 15. d); 16. a); 17. c); 18. b); 19. d); 20. b); 21. c); 22. c); 23. a); 24. b); 25. a); 26. b); 27. d); 28. c); 29. a); 30. b); 31. b); 32. b); 33. d); 34. d); 35. c); 36. c); 37. c); 38. a); 39. a); 40. a); 41. b); 42. b); 43. c); 44. b); 45. b); 46. d); 47. a); 48. c); 49. b); 50. a); 51. c); 52. b); 53. a); 54. b); 55. a); 56. b); 57. c); 58. d); 59. a); 60. a); 61. b); 62. a); 63. a); 64. d); 65. b); 66. d); 67. d); 68. c); 69. a); 70. d).